THE
ROYAL
SOCIETY

THE
ROYAL
SOCIETY

AND THE INVENTION
OF MODERN SCIENCE

ADRIAN TINNISWOOD

BASIC BOOKS
NEW YORK

Basic Books
Hachette Book Group
1290 Avenue of the Americas, New York, NY 10104
www.basicbooks.com

Printed in the United States of America
Originally published in Great Britain in hardcover and ebook by Head of Zeus
in January 2019
First US Edition: June 2019

Published by Basic Books, an imprint of Perseus Books, LLC, a subsidiary of
Hachette Book Group, Inc. The Basic Books name and logo is a trademark of
the Hachette Book Group.

The Hachette Speakers Bureau provides a wide range of authors for speaking events.
To find out more, go to www.hachettespeakersbureau.com or call (866) 376-6591.

The publisher is not responsible for websites (or their content) that are not owned
by the publisher.

Print book interior design by Linda Mark.

Library of Congress Cataloging-in-Publication Data
Names: Tinniswood, Adrian, author.
Title: The Royal Society : and the invention of modern science / Adrian Tinniswood.
Description: New York : Basic Books, [2019] | Includes bibliographical references
    and index.
Identifiers: LCCN 2018034742 (print) | LCCN 2018038569 (ebook) |
    ISBN 9781541673762 (ebook) | ISBN 9781541673588 (hardcover)
Subjects: LCSH: Royal Society (Great Britain)—History. | Science—Great
    Britain—History.
Classification: LCC Q41 (ebook) | LCC Q41 .T56 2019 (print) | DDC
506/.041—dc23
LC record available at https://lccn.loc.gov/2018034742
ISBNs: 978-1-5416-7358-8 (hardcover), 978-1-5416-7376-2 (ebook)

LSC-C

10 9 8 7 6 5 4 3 2 1

For Lucy, with thanks

CONTENTS

PRELUDE

"Ocular inspection"

IMAGINE A UNIVERSE IN WHICH THE SUN REVOLVED around the earth.

In fact, you wouldn't have to imagine it. All you would have to do is believe the evidence of your own eyes. "Isn't it obvious? Every morning the sun rises in the east. It tracks across the sky and then it sets in the west," you might say. "You can see it happening. The very phrase 'sunrise' makes it clear that it is the sun which moves."

And if someone were to suggest to you that it was really the earth that flew through space, and that it spun on its own axis as it did so, you'd be justified in asking how we managed to stand upright. If we were hurtling through the solar system at great speed, why wouldn't there be a perpetual great wind blowing? Why wouldn't we be hurled off into space?

But what if it wasn't only common sense telling you the sun moved around the stationary earth, but that, in a rare display of unity, both Catholic and Protestant theologians were clear on the matter? Imagine the Catholic Church telling us that "the view that the sun stands motionless at the centre of the universe is foolish, philosophically false, and utterly heretical," not least because it contradicted holy scripture: when Joshua asked God for help in his battle against the Amorites, "the sun stood still, and the moon stayed, until the people had avenged themselves upon their enemies." Or Martin Luther poking fun at fools who claimed the earth went around "instead of the sky, the sun, the moon, just as if somebody were moving in a carriage or ship might hold that he was sitting still and at rest while the earth and the trees walked and moved." Or John Calvin saying people were deranged, possessed by the devil, if they really thought the sun was still and it was "the earth which shifts and turns."[1]

If you turned to that other great bastion of authority, the writings of the ancients, you'd receive the same message. Aristotle and Ptolemy, who after all knew rather more about these things than we do, understood perfectly that the universe was geocentric, with the moon, the sun, the planets, and the stars moving around a fixed spherical earth in a series of concentric celestial spheres. Wouldn't you be foolish to think otherwise?

This is the world into which the Royal Society of London was born. A world in which most people, if they thought of cosmology at all, still accepted the Aristotelian and Ptolemaic vision of the universe. True, a few more advanced thinkers were aware that the traditional view was being challenged. They knew of theorists like Nicolaus Copernicus, the Polish astronomer whose *De revolutionibus orbium coelestium* (*On the Revolutions of the Heavenly Spheres*, 1543) was one of the first works to articulate a heliocentric worldview. (The very first was a work, now lost, by the ancient Greek astronomer Aristarchus of Samos, who, according to Archimedes, suggested in the third century BC that the earth revolved around the sun.) They knew of the German astronomer Johannes Kepler, and his theory that the planets moved in elliptical orbits around the sun. But they also knew of Tycho Brahe, who tried to reconcile Ptolemaic and Copernican explanations of the universe by suggesting that, although the five other planets—Mercury, Venus, Mars, Jupiter, and Saturn—did indeed move around the sun, the sun itself moved around the earth. And they had heard of Giordano Bruno, the Italian natural philosopher who, in February 1600, was burned at the stake for defending the Copernican system and maintaining that there were infinite worlds.

In 1633, while most of the founders of the Royal Society were still children (several weren't yet born), Galileo Galilei was hauled before the Inquisition. On pain of death, he recanted his belief that the earth moved, but he was placed under house arrest for the rest of his life.

In medicine, the humoral pathology of the ancient Greek physicians Galen and Hippocrates was still current in the seventeenth century. Medical men believed that good health relied on maintaining a balance among the four humors thought to be mixed together in the veins of the human body: blood was associated with the liver, and phlegm with the brain and lungs; black bile, or *melancholer*, was secreted by the spleen, and yellow bile, or *choler*, by the gall bladder. A marked humoral imbalance was regarded as the cause of most complaints and even dictated character. According to *The Sick-man's Rare Jewel*, a popular medical treatise of 1674, "those in whom phlegm hath the dominion . . . are of a slow capacity, dull, slothful, drowsy, they do dream of rains, snows, floods, swimming and such like."[2] Those of a choleric disposition "are of a quick and nimble wit, stout, hardy and sharp, vindicating of injuries received, liberal even to prodigality, and somewhat desirous of glory, their sleep is light, and that from which they are quickly awaked, their dreams are fiery, burning, quick and full of fury."[3] Most of the population held to these principles of pathology right through to the late seventeenth century, believing in the humors and the ebb and flow of blood from the liver. Galen of Pergamon, a Greek who lived and worked in Rome in the second century AD, had declared that this was how the body worked—even though, as far as we know, he never dissected an adult human being.

In the universities, classical authority underpinned scholarship. Undergraduates in the Faculty of Arts at Oxford had to spend four years "in the study of arts and in diligent attendance, according to the exigence of the statutes, upon the public lectures within the university."[4] This meant grammar, rhetoric, logic, moral philosophy, geometry, and Greek, with a heavy emphasis throughout on Aristotle. In petitioning for his degree, a student stated that his qualifications would "suffice for his admission to lecture on every book of Aristotle's logic." Lectures, which were in Latin, lasted a civilized forty-five minutes, and students were fined for nonattendance. Conversation at dinner and supper was also in Latin. Tutors directed students' studies, read with them each morning, and pointed them in the direction of the right texts: the theologians Bartholomäus Keckermann and Robert Sanderson on logic; Cicero, Pliny, Caesar, and Livy for rhetoric and history; the New Testament in its original Greek; Franciscus Pavonius's *Summa Ethicae* for moral philosophy. A master's degree meant another three years of Aristotle together with Greek, Hebrew, classical history, natural philosophy, geometry, and astronomy. The few texts in the curriculum not written by long-dead authors were commentaries. And they were commentaries on those same long-dead authors.

Outside the universities, and perhaps inside their walls as well, there remained a popular belief in the existence of centaurs, unicorns, and giants; commentators could uphold in all seriousness the notion that serpents were generated from the brains of the dead, that the chameleon lived on air and the ostrich ate iron, that the elixir of youth was a reality, and that basilisks hatched from eggs laid by cocks. Most of the founding members of the Royal Society were born during the reign of James I (1603–1625), a king who firmly believed in magic and witchcraft. The world was a dark, confused place in which any deviation from accepted canons was a dangerous road to travel.

Yet there were challenges to orthodoxy. During a stay in Venice in May 1609, Galileo, then a professor of mathematics at Padua, heard a rumor of a Dutch invention "by the aid of which visible objects, although at a great distance from the eye of the observer, were seen distinctly as if near."[5] Back in Padua, he set about making his own telescopes (he had made four by January

1610, the last and most powerful of which had a magnification of 30x), and when he trained them on the heavens, he was astonished by what he saw. The moon's surface was not smooth, as the Greeks had thought, but covered in mountains and valleys; Jupiter was a round disc with four moons; and, most startling of all, there was a multitude of stars that, invisible as they were to the naked eye, no one had seen before. "Upon whatever part of [the Milky Way] you direct the telescope," wrote Galileo, "straightway a vast crowd of stars presents itself to view; many of them are tolerably large and extremely bright, but the number of small ones is quite beyond determination."[6]

Initial reactions to Galileo's discoveries were mixed: traditional Aristotelians refused point-blank to believe it was possible to see anything in the heavens not mentioned by Aristotle, while governments decided the stars could be left to themselves, seizing instead on the military potential of the telescope. But the ability to observe the previously unobservable universe opened up new worlds—quite literally transforming astronomy and offering unrivaled opportunities for research and wonder, as the English diplomat Sir Henry Wotton was quick to appreciate. In March 1610, Wotton, then James I's ambassador in Venice, dispatched home a copy of Galileo's *Sidereus Nuncius* (*Starry Messenger*), a pamphlet containing work based on his early observations with his new instruments. Wotton included a cover letter:

> I send herewith unto his Majesty the strangest piece of news (as I may justly call it) that he hath ever yet received from any part of the world; which is the annexed book (come abroad this very day) of the Mathematical Professor at Padua, who by the help of an optical instrument (which both enlargeth and approximateth the object) . . . hath discovered four new planets rolling about the sphere of Jupiter [Jupiter's moons], besides many other unknown fixed stars; likewise the true cause of the Via Lactea [Milky Way], so long searched; and lastly that the moon is not spherical, but endued with many prominences, and, which is of all the strangest, illuminated with the solar light by reflection from the body of the earth. . . . So as upon the whole subject he hath first overthrown all former astronomy.[7]

Eighteen years later, William Harvey, the physician to James I and Charles I, published *De motu cordis* (*On the Motion of the Heart*), the product of a decade of research and teaching about the theory that blood circulated around the body instead of ebbing and flowing from the liver. Harvey's critics pointed to Galen to disprove him, arguing that if the blood circulated, all of humoral pathology would be brought into question. If the humors were mixed together, it would be impossible to modify them individually. Others simply asserted that Galen provided them with a working hypothesis, and that was quite enough for them, thank you.

For his part, Harvey—himself a conventional Aristotelian in many ways—couldn't explain why blood circulated—but he was sure that it did, and this was the crucial point. Instead of accepting that classical authorities were right and therefore he must be wrong, as his predecessors had done, he urged the primacy of "ocular inspection." Don't accept things at face value: see for yourself.

He wasn't alone. At the beginning of the seventeenth century the courtier and philosopher Francis Bacon, Viscount St. Alban, had argued for the rejection of traditional Aristotelian learning, suggesting that instead of using unproven hypotheses to test the validity of empirical observations, one should begin with those observations:

> There are and can exist but two ways of investigating and discovering truth. The one hurries on rapidly from the senses and particulars to the most general axioms; and from them as principles and their supposed indisputable truth derives and discovers the intermediate axioms. This is the way now in use.

> The other constructs its axioms from the senses and particulars, by ascending continually and gradually, till it finally arrives at the most general axioms, which is the true but unattempted way.[8]

Bacon's proposal, that learning could be advanced by experiment rather than by refining and interpreting the established authorities, was to prove enormously influential for the young Royal Society. All the founding members were followers of the Baconian method, proponents of experimental learning. Like William Harvey, they stressed the need for "ocular inspection."

Bacon died in 1626. According to tradition, his commitment to experimentation proved to be the death of him. Thomas Hobbes told the antiquary John Aubrey that while Bacon was taking the air in his coach he had a notion to test whether snow would preserve flesh in the same way that salt does. So he stopped his coach and went into the house of a poor woman in Highgate, where he bought a hen from her and asked her to eviscerate it and stuff the body cavity with snow; he even showed her how to do it. It was apparently too much for him: "The snow so chilled him that he immediately fell so extremely ill that he could not return to his lodging," Aubrey wrote.[9] He died two or three days later.

But even in death, Bacon had a contribution to make to the origins of the Royal Society. In his posthumously

published *New Atlantis* (1627), an unfinished fantasy
in which a party of lost travelers comes upon a utopian
island in the South Seas, the philosopher-courtier described Solomon's House, an order of wise men who
devised experiments and sought out reports of those
that had been conducted in foreign countries. The order
boasted extensive chemical laboratories, astronomical
observatories, pharmaceutical and medical facilities, "a
mathematical-house, where are represented all instruments, as well of geometry as astronomy," gardens,
and a repository containing "patterns and samples of
all manner of the more rare and excellent inventions."
Solomon's House members carried out research in optics, microscopy, magnetics, even genetic engineering,
working together to discover "the knowledge of causes,
and secret motions of things; and the enlarging of the
bounds of human empire, to the effecting of all things
possible." This early formulation of the idea of a scientific research institute, a college of like-minded Fellows
who came together to dedicate themselves to promoting
experimental philosophy, was to bear fruit thirty-three
years later in a room at Gresham College in London. It
was there that the Royal Society was born.[10]

1

FOUNDATION

"For the promoting of experimental philosophy"

AT THREE O'CLOCK ON WEDNESDAY, NOVEMBER 28, 1660, the Gresham professor of astronomy, Christopher Wren, finished his weekly lecture in the college reading hall and, still wearing his hood and gown, walked across the courtyard and into the lodgings of his colleague Lawrence Rooke, a professor of geometry.

Besides these two academics, ten other men crowded into Rooke's chambers. They were a motley bunch, a mixture of university teachers and interested amateurs, royalists and republicans, young and old. Rooke had been at Gresham College on Bishopsgate since 1652, when he took up the chair of astronomy there; he switched to geometry in 1657, apparently because the professor of geometry's lodgings were nicer and had a balcony. Young Christopher Wren got the astronomy chair and Rooke's old lodgings in the same year: he was only twenty-five. A third Gresham professor was there, too: Jonathan Goddard, who held the "chair of physic." Other academics included John Wilkins, who, until he was ousted in the aftermath of the Restoration, had been a leading Cromwellian university administrator, warden of Wadham College Oxford, and then master of Trinity College Cambridge; and William Petty, a physician, statistician, and onetime professor of music at Gresham. Petty had risen to fame in Oxford back in 1650. When deputizing at a dissection for the Regius Professor of Medicine, because the latter "could not endure the sight of a bloody body," he had been present at the miraculous resuscitation of a servant, Anne Greene, who had been hanged in Oxford Castle for murdering her illegitimate child. Greene had been carried to Petty's lodgings for dissection, and while he prepared his instruments, a spectator—dissections being rather public affairs in the seventeenth century—noticed she was still breathing and stamped on her chest, thus unwittingly performing cardiopulmonary resuscitation and enabling Petty to revive her. She was soon well enough to go home, taking her coffin with her as a memento of her miraculous preservation.[1]

After the academics, the second-largest group at Rooke's consisted of four professional courtiers: Sir Paul Neile, an amateur astronomer; Sir Robert Moray and Alexander Bruce, two old soldiers and supporters of the Stuart cause in exile, who had returned to England when Charles II was restored to the throne; and William, Viscount Brouncker, who had kept his head down during the 1650s but emerged as an eager royalist when it was safe to do so. Unlike the other three, Brouncker had an established scientific reputation—he was a mathematician.

Finally, three men were present who belonged to neither court nor college. William Balle, who had lodgings in the Temple, was an amateur astronomer; in the 1650s, along with Neile and Wren, he had become interested in Saturn and its changing outline (telescopes were advanced enough to show that the planet changed its shape, but not yet good enough to show the reason— it was surrounded by rings that presented themselves in different alignments at different times to observers on earth). Abraham Hill, at twenty-five the youngest member of the group (Moray, at fifty-one or fifty-two, was the oldest), was a London merchant whose parents had just died, leaving him a considerable legacy: his presence at Rooke's lodgings was probably due to the fact that with his new fortune he had rented chambers of his own at Gresham College. The third and last of this group was Robert Boyle, the brilliant, neurasthenic son of the Earl of Cork who just that year had published *New Experiments Physico-Mechanical, Touching the Spring of the Air and Its Effects*. The book described how, together with his assistant, twenty-five-year-old Robert Hooke, he had constructed a vacuum chamber and demonstrated the effects of the withdrawal of air on flame, light, and life itself. Boyle was something of a maverick—reclusive, fiercely Protestant in a country that was violently divided on the issue, usually unwilling to join any club that might have him as a member.

These friends, and friends of friends, spoke on that day, as they usually did when they met, of scientific matters and new discoveries. They shared ideas and theories and discussed experimental philosophy. These discussions led to the first talk of forming a more formal association:

> And because they had these frequent occasions of meeting with one another, it was proposed, that some course might be thought of to improve this meeting to a more regular way of debating things; and that, according to the manner in other countries, where there were voluntary associations of men into academies for the advancement of various parts of learning, they might do something answerable here for the promoting of experimental philosophy.[2]

4

So they agreed to hold regular weekly meetings on Wednesdays at three o'clock. During the term they would meet in Rooke's lodgings; William Balle offered his rooms in the Temple for weeks when Gresham was not in session. Everyone would pay ten shillings as a one-off admission fee and a weekly subscription of one shilling, whether they came to a meeting or not. John Wilkins was appointed as chair, with Rooke as treasurer. The Gresham professor of rhetoric, a physician named William Croone, was made registrar, or secretary, even though he wasn't actually present. And with a view to expanding their membership, the twelve put together a list of forty prospective recruits, "such persons, as were known to those present, and judged by them willing and fit to be joined with them in their design."[3]

The Royal Society was born.[*]

❦

If the date of the Society's birthday is pretty clear, its parentage is still a matter for heated debate, and historians of science—who, like the seventeenth-century figures they write about, are driven to seek for precedence and priority—argue over its ancestry. Did the

---

[*] Further details on the lives of the founders can be found in Appendix 1, pp. 133–149.

Society grow out of an "invisible college," a loose grouping of experimental philosophers mentioned by Robert Boyle in 1646 and 1647? Was it an offshoot of the "Great Club" based at Wadham College, which was described by Seth Ward, Savilian Professor of Astronomy at Oxford, in 1652?* Or did it emerge from an accidental coming-together of academics and like-minded philosophers who drifted into London and began to congregate at Gresham College in the last years of the Commonwealth?

The answer is yes to all three. The Oxford mathematician John Wallis, who was on the list of prospective members drawn up after that inaugural meeting, remembered how, in 1645, when the Civil Wars between the Royalists and the Parliamentarians were still raging, a group of enthusiasts would sometimes meet in Jonathan Goddard's London lodgings, and sometimes at the Mitre in Wood Street, in the heart of the city. Besides Goddard and Wallis, the group included John Wilkins, who had published a string of popular scientific works: *The Discovery of a New World* (1638), which speculated on the possibility that the moon was habitable; *A Discourse Concerning a New Planet* (1640), which argued for the cosmological vision propounded by Copernicus,

---

*In 1619 the mathematician Sir Henry Savile, warden of Merton College, founded two chairs, in astronomy and geometry, that still bear his name today.

Kepler, and Galileo; and *Mercury, or The Secret and Swift Messenger* (1640), on the use of codes and ciphers. Most of the other members of this group were medical men; prominent among them was Charles Scarburgh, who had a successful and fashionable medical practice but pursued his interests in mathematics and optics as well. According to his near-contemporary Walter Pope, Scarburgh "lived magnificently, his table being always accessible to all learned men, but more particularly to the distressed Royalists, and yet more particularly to the scholars ejected out of either of the universities for adhering to the King's cause."[4] As a teenager, Christopher Wren lodged with him; later Wren gave the older man credit for fostering his interest in the mathematical sciences.

Wallis, who admittedly was writing twenty-five years after the events he described, was certain that these London gatherings were the ancestors of the Royal Society. Members paid a weekly contribution toward the charges incurred in preparing experiments, he claimed, and adhered to a set of rules. Talk of religion and politics was forbidden. The group confined itself to "physic, anatomy, geometry, astronomy, navigation, statics, mechanics and natural experiments."[5] They discussed the circulation of the blood (William Harvey was a friend of Scarburgh's and may have been present at some of the meetings), the Copernican hypothesis and the nature of comets, the acceleration of

7

bodies through the air, and improvements in optics. These meetings, Wallis went on to say, moved later to the Bull Head Tavern in Cheapside and Gresham College. "And our numbers increased," he reported.[6]

The weekly meetings Wallis described petered out in about 1648. It is no coincidence that it was in April of that year that John Wilkins, who seems to have been a prime mover in all the precursors to the Royal Society, moved to Oxford, where he was appointed warden of Wadham College in the sweeping reforms that saw hundreds of Royalists expelled from the university, and the heads of most colleges replaced by men who were sympathetic to the Parliamentary cause. Wilkins quickly established an open and tolerant regime at Wadham, with "nothing of bigotry, unmannerliness, or censoriousness, which then were in the zenith amongst some of the heads and fellows of colleges in Oxford," according to Walter Pope.[7] He encouraged experimental philosophy, and very soon Robert Boyle, John Wallis, Jonathan Goddard, and others had moved to Oxford to attend the scientific meetings that Wilkins established as a continuation of the invisible college in London. They were joined by several Cambridge men, including Lawrence Rooke and Rooke's mentor, Seth Ward, a friend of Scarburgh's who had been deprived of his fellowship at Cambridge for opposing the Solemn League and Covenant of 1643.* Ward was appointed Savilian Professor of Astronomy under Oxford's slightly more tolerant regime. The constellation of talent, and Wilkins's nonpartisan regime, began to attract gifted undergraduates from both sides of the political divide. Christopher Wren, whose Royalist uncle was currently imprisoned in the Tower for his opposition to Parliament, arrived at Wadham in 1649 and became part of the circle.

In 1652 Seth Ward described a "Great Club" of about thirty members who had set themselves the task of recording "such things as are already discovered," and then making a list of what still needed to be discovered and devising relevant experiments. About eight people had also joined together, he wrote to a friend, "for the furnishing [of] a laboratory and for making chemical experiments which we do constantly every one of us." Ward was himself building an observatory on the roof of the gatehouse at Wadham and procuring telescopes "and other instruments for observation."[8]

The following year, Samuel Hartlib, a Polish émigré who maintained an extensive correspondence network

---

*The Solemn League and Covenant was an alliance between the English Parliament and the Scots whereby Parliament received military aid in its fight against Charles I's armies in return for a commitment to the Presbyterian system of church government favored by the Scots.

and took a lively interest in anything to do with the new philosophy, heard that Wilkins had established "a college for experiments and mechanics at Oxford," and he contributed £200 of his own money toward it.[9] When the diarist John Evelyn visited Oxford in 1654, he was invited to dinner by "that most obliging and universally-curious" Wilkins and saw some of the results of the "experiments and mechanics":

> He . . . showed me the transparent apiaries, which he had built like castles and palaces, and so ordered them one upon another, as to take the honey without destroying the bees. These were adorned with a variety of dials, little statues, vanes, &c. . . . He had also contrived a hollow statue, which gave a voice and uttered words by a long concealed pipe that went to its mouth, whilst one speaks through it at a good distance. He had, above in his lodgings and gallery, variety of shadows, dials, perspectives, and many other artificial, mathematical, and magical curiosities, a way-wiser [an instrument for measuring distances traveled], a thermometer, a monstrous magnet, conic, and other sections, a balance on a demi-circle, most of them of his own, and that prodigious young scholar Mr. Christopher Wren.[10]

Robert Boyle, though never directly involved with the university, had rooms on High Street, where he and Robert Hooke carried out experiments with the air pump. Wren performed the first successful canine splenectomy in the world in Boyle's rooms, using a sow-gelder's knife on Boyle's spaniel. William Petty was in Oxford, carrying out dissections for the squeamish Regius Professor of Medicine; the group met in his lodgings for a time because he was staying in an apothecary's house, which meant that drugs, chemicals, and other items were on hand. So was John Wallis, who was appointed Savilian Professor of Geometry in June 1649, a post he held for the next fifty-four years; and the pioneering physician and chemist Thomas Willis, who had a house on Merton Lane, lived just around the corner from Boyle. Seth Ward taught Copernicus, the first Savilian Professor to do so. Under Wilkins's aegis, experimental philosophy flourished in Oxford as never before.

As is the nature of such things, the group dispersed over time, and some of its members lost interest. Petty went to Ireland as Cromwell's physician-general in 1652. Wren got the chair in astronomy at Gresham in 1657; Rooke was already there, and Goddard followed Wren in 1658. When John Wilkins moved to Cambridge in 1659 as master of Trinity College, the impetus behind the so-called Oxford Experimental Philosophical Club was already fading.

Or perhaps it would be fairer to say it was shifting. Gresham College, which already had a distinguished reputation as a cradle of the sciences, was another of the Royal Society's progenitors. It had been founded in 1597 under the terms of the will of Sir Thomas Gresham, a wealthy Elizabethan merchant whose other claim to fame was that he'd founded London's main commercial center, the Royal Exchange, in Cornhill. Gresham left his town house, a sprawling courtyard mansion that lay between Bishopsgate and Broad Streets, as a sort of adult education center for the capital. In return for fifty pounds a year and free lodgings in the college, professors of divinity, law, physic, rhetoric, music, geometry, and astronomy gave open weekly lectures to the citizens of London in the great hall behind the main block. The chairs in geometry and astronomy were the first to be established in England, and during the first three decades of the seventeenth century Gresham College developed a considerable reputation as a center for research in the mathematical sciences. It attracted men of the stature of Henry Briggs (professor of geometry from 1597 to 1620), who was described by one of his contemporaries as "the English Archimedes"; and Edmund Gunter (professor of astronomy from 1619 to 1626), whom John Aubrey claimed was "the first that brought mathematical instruments to perfection."[11] Both men had strong Puritan sympathies and particular interests in navigation and shipping, and several of their successors shared both their Puritanism and their research interests. A close collaboration grew up with the instrument-makers, naval architects, and artisans at the dockyards in Deptford, which turned the college into a prominent center for applied mathematics and navigation.

During the 1630s and 1640s, however, Gresham's reputation declined from the peak it had reached under Briggs and Gunter. Far from being at the leading edge of technological advancement, it became the subject of complaints from students. The professors were failing even to fulfill their less-than-arduous teaching duties, either paying deputies to read for them or not bothering to lecture at all. The professor of divinity, Richard Holdsworth, spent most of his time at Cambridge, where he was also master of Emmanuel College; the professor of law, Thomas Eden, had to resign his chair in 1640 because of "several other employments . . . interfering with his attendance at Gresham"; the professor of geometry, John Greaves, went off to explore the Middle East in 1633 and didn't come back for seven years (he was eventually sacked "on account of his long absence, and his neglect of his lecture").[12]

In 1651 William Petty became the unlikely professor of music at the college, although his duties in

Ireland meant that he was rarely there. A second member of the Oxford group, Lawrence Rooke, was elected professor of astronomy in 1652; Jonathan Goddard became professor of physic, "through the favour and power of Cromwell," three years later; and in August 1657, Christopher Wren took up the vacant chair in astronomy when Rooke moved to geometry and those comfortable lodgings with the balcony.[13] In the space of six years, four of the seven professorships, including all three of the scientific chairs, belonged to members of the group that had gathered around Wilkins at Oxford.

The Restoration of Charles II as monarch and head of state in 1660 meant the return to London of Royalist exiles such as Alexander Bruce and Sir Robert Moray. It meant that lukewarm Royalists like Brouncker and Sir Paul Neile were there, too, proclaiming their loyalty and looking for preferment from the new regime. And ardent Cromwellians like Wilkins and Petty found themselves out of jobs and with time on their hands. The conditions were right for Solomon's House. And on that Wednesday afternoon in November 1660, the foundations were laid and the work began in earnest.

2

CHARTER

"Our philosophic assembly"

AFTER THAT FIRST MEETING IN ROOKE'S LODGINGS at Gresham College the group convened again on Wednesday, December 5, 1660, as arranged. The courtier Sir Robert Moray, who lived at the Palace of Whitehall, reported that Charles II had been told about the project "and well approved of it."[1] Now the members set about creating an institutional framework, a constitution that would ensure that their society, which still didn't have an official name, would have an existence and objectives above and beyond those of individual members. Its aim, they decided, was to meet weekly "to

consult and debate concerning the promoting of experimental learning."[2]

Things moved quickly. At the third meeting, on December 12, after a committee had been formed to create a draft constitution, the members voted on a whole
raft of rules and regulations. They set a limit of fifty-five
members, with a quorum of twenty for elections to office, and agreed that no one should be admitted "without
scrutiny." The only exceptions to this were Fellows of
the Royal College of Physicians; professors of mathematics, physic, and natural philosophy at Oxford and
Cambridge; and "such as were of, or above, the degree
of baron."[3] The value of attracting influential aristocrats
into the new organization was obvious. Money and
proximity to power were high on the society's agenda,
and aristocrats could promise both, although in practice
few of them delivered. Moreover, the decision to admit
noblemen without reference to their scientific credentials
set the Royal Society on a path to a mixed membership
that included men with social status and rank, but only a
passing interest in science. Aristocrats made up around
20 percent of the membership in the early years, with
courtiers and politicians (as distinct from nobles) making up another 25 percent. Some were active members;
others never attended a single meeting. They did, however, provide the young Royal Society with a certain
social cachet. In 1670, by which time Charles II, his

brother James, Duke of York, and their cousin Prince Rupert were all members, the Swedish scholar Georg Stiernhielm wrote in awe that "like a second Apollo the king himself presides as supreme moderator and governor of this band of stars, among whom are to be found the sons of kings, princes, dukes, magnates, landowners, counts, barons, great patrons of learned men, and a host of men of all orders distinguished for their learning and wisdom."[4]*

The members confirmed the appointment of the three main officers—a president, or director, who was to be chosen monthly, a treasurer, and a register—or, as we would say today, a "registrar." These last two were to continue in office for a year. From the beginning, the group recognized the importance of keeping written records: the register was to provide one book for the statutes of the society and the names of its members, another to record experiments, and a third "for occasional orders."[5] He was assisted by an amanuensis, who was paid forty shillings a year. There was also an operator, who for four pounds a year assisted with experiments and demonstrations.

Lastly, the group established a method of electing members. No one could be elected on the same day he

---

*The royal connection continues to this day: Queen Elizabeth II is a Fellow, as well as being patron of the Royal Society, and Prince Charles is also a Fellow.

was proposed, and a two-thirds majority of those present was needed for the candidate to be admitted. The actual voting was conducted as follows:

> The amanuensis [should] provide several little scrolls of paper of an equal length and breadth, in number double to the members present. One half of these to be marked with a cross, and the other with cyphers; and both being rolled up, to be laid in two distinct heaps. Every person then coming in his order shall take from each heap a roll, and throw which he shall please privately into an urn, and the other into a box. After which the director, and two others of the society, having openly numbered the crossed rolls in the urn, shall accordingly pronounce the election.[6]

The secret ballot might not seem particularly advanced today, but we should remember that it would be more than two hundred years before the Ballot Act introduced secret ballots for parliamentary and municipal elections in Britain. The Society was far ahead of its time.

A constitution gave structure to the new institution; but for real stability and the assurance of continuity, incorporation by royal charter was the only route, since it transformed the institution from a collection of individuals into a single and lasting legal entity. Cities and universities had royal charters; so did the universities, the London livery companies, and the big overseas trading companies. Charles I had granted a dozen royal charters to organizations ranging from the Playing Card Makers' Company to an almshouse. In the first two years after Charles II's restoration to the throne, three new royal charters were granted: to the Glass Sellers' Company, the Royal African Company, and the New England Company. A royal charter brought a measure of protection; it also brought recognition, legitimacy, and validation.

While Sir Robert Moray lobbied the king for a charter, members searched for a suitable name for the new institution. At one point it was being called an academy, perhaps in reference to the Académie Montmor, a group of natural philosophers who met weekly at the Hôtel de Montmor in Paris in the late 1650s and early 1660s. In July 1661, the lawyer John Hoskins wrote to John Aubrey to say that "I wonder why you tell me nothing of the famous Academy of our philosophical sceptics that believe nothing not tried."[7]

Credit for the name "Royal Society" went to John Evelyn, who, in a fawning panegyric to mark Charles II's coronation on April 23, 1661, praised the king for encouraging the group's efforts, writing of "the honour you have done our Society at Gresham College." Seven months later, in the dedication to his translation of Gabriel Naudé's instructions on setting up a library,

Evelyn thanked Lord Chancellor Clarendon for "the promoting and encouraging of the Royal Society." On December 3, 1661, he noted that, "by universal suffrage of our philosophic assembly, an order was made and registered that I should receive their public thanks for the honourable mention I made of them by the name of Royal Society."[8]

With the name gaining acceptance, Sir Robert Moray was working hard behind the scenes to procure a royal charter. Moray, who served as president nine times between March 1661 and July 1662, more than any other member, was a Scottish privy councillor with his own set of lodgings in the Palace of Whitehall and easy access to Charles II. It was he who brought the news of the king's approval in the days following the first meeting in Rooke's lodgings at Gresham, and in October 1661 he reported that he and Sir Paul Neile "had kissed the king's hand, in the society's name," when they presented him with a petition requesting a charter of their own.[9] At the beginning of June 1662, Moray was appointed to a committee "to draw up a paper concerning the design of the society," along with Lord Brouncker, William Petty, Jonathan Goddard, and John Wilkins.[10] On July 15, 1662, his efforts bore fruit when the Royal Society's charter was formally authorized by Charles II. The following month, Lord Brouncker and as many members as he could muster waited on the king at Whitehall to thank him for his generosity and interest in the Society. They assured him of their "firm resolution to pursue sincerely and unanimously . . . the advancement of the knowledge of natural things, and all useful arts, by experiments."[11] They also thanked the lord chancellor, along with Sir Robert Moray "for his concern and care in promoting the constitution of the society into a corporation."[12]

Under the terms of the charter, Lord Brouncker had been appointed president of the Royal Society rather than Sir Robert Moray or the other obvious contender, John Wilkins. The reasons aren't altogether clear, unless it was that a nobleman seemed a more appropriate choice, or that Brouncker's other role, as chancellor to Charles II's new queen, Catherine of Braganza, brought with it more influence at court. The charter stipulated that the president was to serve until the feast of St. Andrew, November 30, after which he could stand for election annually. The ruling body was to be a council of twenty-one, including the president; Moray, Boyle, Wallis, Neile, Wren, and Petty were among those named. They were also to serve until St. Andrew's Day, when ten should stand down to allow others to take their place, a process that was to be repeated every year. William Balle was appointed treasurer in place of Lawrence Rooke, who had died that June; John Wilkins and the German émigré Henry Oldenburg were the two secretaries. Members were henceforth to be known as Fellows of the Royal Society, an echo, perhaps, of Bacon's *New Atlantis*, in which the members of Solomon's House are referred to as Fellows.

Along with the promise of continuity as a single legal entity and the security of an organizational structure enshrined in law that the charter brought, it came with several other advantages. The Society was licensed to hold a correspondence "on philosophical, mathematical, or mechanical subjects, with all sorts of foreigners"; even more importantly, it was allowed to authorize individual publications—a huge privilege in an age when all forms of media were strictly regulated by the state. The Society was also granted the right "to require, take, and receive the bodies of such persons as have suffered death by the hand of the executioner, and to anatomize them."[13]

Delighted though they were with their new status, the Fellows immediately began to lobby for more. Meetings of the council were put on hold, as were the elections on November 30, 1662. Moray, Neile, and Brouncker went to work again, and the result was another royal charter, which the king authorized on April 22, 1663.

This second charter emphasized that the Royal Society of London for Improving Natural Knowledge, as it was now called, had close links with the crown.

Charles II now declared himself founder and patron—
they had high hopes that Charles would show his ap-
preciation for their efforts by making a large donation
or grants of land. The charter also provided for a dramatic expansion in numbers from the original target of
fifty-five: anyone the president and council admitted
into the Society within two months from the date of
the charter was declared to be a member. At a meeting
on May 20, 1663, a total of 115 men were admitted as
Fellows of the Royal Society. Comprising both existing
members and new recruits, they included a smattering
of earls, nineteen knights, and twenty-two medical
men. By 1671 two out of three scientific Fellows—that
is, Fellows with a professional interest in one or more
branches of the sciences—were physicians or surgeons,
a proportion explaining the fact that inquiries into
physiology and anatomy dominated many of the Society's early meetings. Five more Fellows were elected
on June 22, including a herald, a journalist, a Scottish
politician, and two foreigners: the distinguished Dutch
scientist Christiaan Huygens, who had been in correspondence with Moray for several years over his theories about the rings of Saturn, among other things; and
Samuel de Sorbière, a French physician and secretary
to the Académie Montmor, who was visiting England
at the time and contributed to several of the Society's

meetings that summer, with a view to establishing a regular correspondence between the two groups.

The Society has always welcomed foreign members. The Polish astronomer Johannes Hevelius was admitted in 1664, the French mathematical scientist Pierre Petit in 1667, the German philosopher Gottfried Leibniz in 1673, and the pioneering Dutch microbiologist Antonie van Leeuwenhoek in 1689. In fact, one in ten of all the Fellows elected between 1660 and 1685 were foreign.

The foreign members rarely attended meetings: many of them were elected as a means of recognizing their scientific achievements and extending the prestige of the Society overseas. Others just happened to be in the right place at the right time. For example, the Moroccan ambassador Mohammed ben Hadou, who arrived at the court of Charles II in December 1681 and became something of a celebrity in London, expressed an interest in the work of the Society and was invited to one of its meetings, where he was shown "a very fair Alcoran [Koran] written in Arabic."[14] Before he left, he added his name, in Arabic, to the list of Fellows.

Foreign members were exempt from paying subscriptions (but then rather too many English members regarded themselves as being in this category, too). After 1682 they formed a separate category on the printed lists of Fellows, and by 1740, the names on the Foreign

List made up nearly half of the total membership of the Royal Society.

∾

The Society was granted a third royal charter in 1669 that provided the hoped-for royal grant of lands, in the form of Chelsea College. The college was a half-built academy for the clergy; it had been founded by James I and used as a prisoner-of-war camp for Dutch sailors during the Commonwealth. The Society planned to turn the place into a Solomon's House, a research institution that would be a permanent home, but there was never enough money for the project. It was sold back to Charles II in 1682 for £1,300 as a site for a new Royal Hospital for veteran soldiers.*

The second charter of 1663 had granted the Royal Society the right to its own arms, a shield bearing the three lions of England, with two white hounds for supporters and a crest of a helm surmounted by an eagle. John Evelyn, who enjoyed this sort of thing, played with various more obviously scientific alternatives, such as a hand issuing from a cloud and grasping a plumb

---

*A fourth charter was granted in 2012: known as the Supplemental Charter and still in use, it made a number of minor amendments to the previous documents, including the adoption of an English translation of the second and third charters, which was to take precedence over the original Latin text.

line, and a terrestrial globe beneath a human eye, but he seems to have been overruled. Charles II presented the Society with a silver-gilt mace, and the arms were promptly engraved on it; Evelyn's father-in-law gave them a cushion to rest it on. No meeting of the Royal Society could or can be held even today without the presence of that mace.

Evelyn had better luck with his ideas for a suitable motto. Along with Latin tags that translate as "How much we do not know," "By experience," and "Try all things," he came up with *Nullius in verba*, which roughly translates as "Take no one's word for it." This was the motto that the Society adopted, and it remains its motto to this day.

3

EXPERIMENT

*"The key that opens treasures
is often plain and rusty"*

INCORPORATED, ARMIGEROUS, AND DULY CONSTI-
tuted, the Royal Society of London was ready to take
its place in the world.

But what exactly *was* that place? What did the
Royal Society *do*?

It brought together knowledge, certainly. From
the beginning, Fellows were asked to prepare narrative
accounts of different trades and industries—engraving
and etching, refining, shipbuilding, masonry, brewing.
The physician Christopher Merrett produced an ambitious catalog "of the natural things of England, and

of the rarities thereof."[1] This objective wasn't pursued merely from a desire to acquire knowledge for its own sake: as the membership grew, there were dabblers and earnest country squires who delighted in curiosities and oddities, but the men at the core of the Society were moved by the conviction that the improvement of natural knowledge would necessarily lead to improvements in trade, commerce, and manufacturing.*

But making "faithful records of all the works of nature, or art," as Thomas Sprat wrote in his 1667 *History of the Royal Society*, "so the present age, and posterity, may be able to put a mark on the errors which have been strengthened by long prescription," wasn't enough.[2] In the words of an undated set of proposals that were probably drawn up by William Neile, son of Sir Paul Neile, "the business of the society is to make experiments"—although not for their own sake: Neile went on to stress that "the experiments themselves are but a dry entertainment without the indagation [i.e., searching out] of causes."[3] At the second meeting of the group, on December 5, 1660, Christopher Wren was asked "to prepare against the next meeting for the pendulum experiment," and a committee consisting of Brouncker,

*Merrett is also credited with inventing the method for making champagne. In a paper on wine-making that he gave to the Society in 1662, fifty years before the process was first mentioned in France, he described how sugar could be added to wine to produce a second fermentation and make it sparkle.

Boyle, Moray, Petty, and Wren was asked "to prepare some questions, in order to the trial of the quicksilver experiment upon Tenerife."[4] The first of these two experiments presumably related to Wren's research into the vibrations of a pendulum, which was thought to hold the key to accurate time-keeping, and hence the solution to the problem of longitude; the other was a reference to plans for observing the behavior of a mercury barometer at the top of a peak in Tenerife, the largest of the Canary Islands, which reached a height of about 3,660 meters (12,000 feet). The Royal Society was many things: a forum for sharing ideas and discoveries, a hub for gathering and disseminating those ideas, and a center for coordinating national and international research. But the demonstration, sponsorship, discussion, and promotion of experimental learning was at its heart.

At the fourth meeting, on December 19, every member was requested to bring in "such experiments, as he should think fit for the advancement of the general design of the society."[5] By January 2, 1661, business was in full swing. Brouncker and Boyle had come up with a list of twenty-two ideas of things to try in Tenerife, including: "observe accurately the time of the sun's rising on the top of the hill and below; and note the difference"; and "try by an hour-glass, whether a pendulum clock goes faster or slower on the top of the hill than below."[6] The list was entered into the register, as a permanent record and as evidence of priority should its contents lead to any important discoveries. Boyle was asked to show his air pump, Jonathan Goddard his "experiments of colours," and Petty the "diagrams of what he had discoursed to the society that day, and the history of building of ships."[7] Brouncker was asked to "prosecute the experiment of the recoiling of guns, and to bring it in at the next meeting." He didn't, and had to be asked again a few weeks later, and again after that—a pattern of promise, nonperformance, and reminder that was to become familiar to many members, especially those with busy professional lives outside the weekly meetings at Gresham. The phrase "bring in" was ambiguous: sometimes it meant an actual demonstration, sometimes a paper reporting on the results of an experiment carried out elsewhere. The request for Goddard to bring in his "experiments of colours," for example, resulted in his producing a paper describing how he mixed different liquids that were either colorless or had a color completely different from that produced when they were brought together. The paper was duly written up in the register.

But plenty of actual experiments were performed at the weekly meetings—around two dozen during the first year. They often resulted in the death of the subject. At the June 13, 1661, meeting, the physician Walter

Charleton, who was researching animal physiology, administered three grains of nux vomica (strychnine) to a young thrush and a young woodpecker, and both died. "He gave also to another young thrush two grains of nux vomica and as much sublimate mixed together; and that killed the bird in nine minutes' time," according to a record of the meeting. Other members also pursued an interest in poisons. A toad and a slowworm were killed by having salt thrown on them, and "it was ordered that two puppies be provided against the next meeting, for the trial of experiments of poisons" (unusually, they both survived). Various other creatures died in Boyle's air pump experiments.[8]

Boyle and some of the others were also prompted to carry out research into buoyancy after Charles II, in an early mark of favor, gave Sir Paul Neile five little glass bubbles "in order to have the judgement of the society concerning them."[9] Boyle brought in three glass cylinders filled with water, with empty glass bubbles inside them, "some of which ascended, the water being heated, and others descended by the like heating; and others ascended, the water being cooled in cold water, and descended by heating the water." Clearly these experiments did not always yield conclusive results.

As the Society's reputation grew, reports began to come in of experiments carried out elsewhere by nonmembers. As a rule, members repeated such experiments to verify their findings, although their own reports were usually accepted without question by Fellows. Members and their guests also brought curiosities to the weekly meetings in an early form of show-and-tell. Some were more curious than others: one member produced a flat stone found at the seaside, because it looked like a biscuit cake. On another occasion the same member, the apothecary John Houghton, proudly presented a four-legged chicken from Godalming. He also showed a yam, half a large bladder thought to be the crop of a pelican, and "a curious humming-bird," which presumably was dead.[10] Dr. Croone brought in a dead parakeet to be embalmed by the botanist and physician Nehemiah Grew, but Grew was absent from the meeting; Brouncker had to take the thing away and hand it over to him later. When he finally got it, Grew decided it wasn't fit to be preserved, promising an embalmed eagle instead.

Sometimes the records were bewilderingly brief: "Dr Wilkins produced a piece of tinged hanging."[11] "The amanuensis was ordered to bring in a glass-hatband at the next meeting."[12] The Duke of Buckingham, who was admitted in June 1661, promised to bring in a piece of a unicorn's horn. There is no record of him ever doing so, but Sir Robert Southwell later produced "a great horn, said to be an unicorn's," and at a meeting in July 1661 a circle was made with powder of unicorn's horn.

When a spider was placed in the middle of it, "it immediately ran out."[13]

Some experiments were harder to do than others. Mathematical demonstrations were considered boring. Those involving ballistics, altitude, or astronomy had to be conducted elsewhere for practical reasons. Vivisections, which would have to be performed without benefit of anesthetics, were unpleasant and often distressing, even to hardened seventeenth-century sensibilities. And specialized equipment was often required for the simplest experiments. That was one of the advantages of meeting in Rooke's rooms at Gresham: he had access to the college's instruments. At one of the very first meetings, the amanuensis noted, "Mr Rooke to provide tubes and quicksilver for the quicksilver experiment."[14]

In the summer of 1663, when the Society was gearing up for a projected visit from Charles II (a visit that never happened), there was much talk about a program of experiments that would amaze, amuse, and entertain its royal patron. But what to choose? Christopher Wren wrote a long and practical letter of advice. Chemical experiments would be dirty and tedious, he thought. Anatomical demonstrations were "sordid and noisome." Mathematical proofs and the display of astronomical instruments were frankly incomprehensible to the uninitiated. Agricultural and industrial machinery needed "letters and references," and more time than a royal visit allowed. "Scenographical, catoptrical, and dioptical tricks" required more skill in the execution than the Society's resources would permit. Architectural designs were no use without an actual building, or at least more insight into the subject than the king could be expected to possess. Surprise and spectacle were what was needed, not conjuring tricks, which would devalue the work of the Society. To "produce knacks only . . . will scarce become the gravity of the occasion"; yet at the same time, "the key that opens treasures is often plain and rusty; but unless it be gilt, the key alone will make no shew at court." Wren went on to offer some practical suggestions: a circular barometer, an artificial eye (i.e., a model of an eye), or a compass suspended in water and resting on springs, so that it could be used in a moving carriage.[15]

With the granting of the first charter, the Fellows had begun to think about professionalizing their approach to experimental philosophy, or at least placing it in the hands of someone with a responsibility for doing it, rather than leaving it up to a dozen or so individual Fellows, most of whom had busy lives beyond the confines of Gresham College. In November 1662, Sir Robert Moray suggested that they find "a person willing to be employed as a curator by the Society, and offering to furnish them every day, on which they met, with three or four considerable experiments."[16] This curator of experiments would need an alternative source of income: Moray was clear that no pay would be forthcoming until the Society was on a sounder financial footing.

The courtier already had someone in mind. Robert Boyle's assistant, Robert Hooke, was familiar with those members who had been in Oxford in the 1650s. After an unsuccessful attempt at being an artist (he was briefly apprenticed to Sir Peter Lely) and a spell as a pupil at Westminster School, Hooke had gone up to Christ Church in 1653 or 1654, where he had assisted first Thomas Willis and then Robert Boyle in their experiments. As a result, he was in frequent contact with the Wadham group—John Wilkins, Seth Ward, John Wallis, and Christopher Wren. He was still working as Boyle's assistant when the Society was founded, and it is likely that he helped Boyle with the experiments he demonstrated at the weekly meetings. His first piece of published scientific research, a tract that sought to explain capillary action, had been entered for debate at the meeting of April 10, 1661.

So when Moray suggested that the Society take on a curator of experiments, everyone knew he meant Hooke. There was unanimous agreement on this choice, and Hooke was appointed the following week. It was ordered that he should "come and sit amongst them, and both bring in every day of the meeting three or four experiments of his own, and take care of such others, as should be mentioned to him by the society."[17] Boyle was thanked for giving up his assistant, and with good reason, since he seems to have carried on paying Hooke's wages.

Of the men who shaped the Royal Society in its early years, John Wilkins had the vision the new organization required. It had been forged during those gatherings in London and Oxford in the Civil War and its aftermath. Sir Robert Moray deserves more credit than he usually receives for his role in gaining royal approval for the project and the legitimacy that the charters brought. But it was Robert Hooke whose commitment to the rigorous practice of experimental philosophy maintained the Royal Society in the vanguard of the scientific revolution during the last third of the seventeenth century. Brilliant, irascible, vain, with a private life that was not without its dark side,* he presented, devised, reported on, and published experiments in almost every branch of science. We're told that "he made but a mean appearance, being short of stature, very crooked, pale, lean, and of a meagre aspect, with lank brown hair."[18] He could be tremendously difficult in his

---

*Always a meticulous recorder of events, Hooke made a careful note in his diary whenever he achieved orgasm, sometimes by himself, sometimes with one of a succession of maidservants, and occasionally with his teenage niece and ward, Grace.

relations with others. When he felt he wasn't being valued enough, he would threaten to leave the Society and set up an alternative group, but he never did. At other times, when his propensity to imagine slights tipped over into outright paranoia, he would rage against his colleagues and swear he would have no more to do with them, only to be drawn back into the powerful scientific circle he had helped to create. But he was a genius, and along with Boyle and Wren, the most gifted of the early Fellows of the Royal Society until Isaac Newton arrived on the scene.

From the beginning, Hooke's status was problematic. Although he was an employee, he ranked as something more than an artisan or tradesman, a point the Society recognized in May 1663 by electing him a Fellow. All the same, members routinely "ordered" him to carry out this experiment or that. The resignation from Gresham College in 1664 of Isaac Barrow, who had succeeded Lawrence Rooke as Gresham professor of geometry, left the chair vacant, and Hooke was an obvious choice to fill it. If he got the post, the set of lodgings and the salary of fifty pounds a year that went with it would allow the Society to carry on having his services for nothing. In the event, the chair went to the physician Arthur Dacres—but a philanthropic city financier, Sir John Cutler, offered to endow a lectureship at Gresham especially for Hooke. Cutler was rather slow to pay Hooke the promised fifty pounds, and the relationship ended in court. Dacres, however, was forced out after only ten months. Hooke then took his place, and he remained at Gresham as professor of geometry until his death in 1703.

In 1664 the Society at last began to pay Hooke, although they docked fifty of his agreed eighty-pound salary because Cutler was deemed—wrongly, as it turned out—to be making up the difference. Nevertheless, the payments made Robert Hooke the first professional research scientist.

Within days of his initial (and unpaid) appointment, he was getting down to work. At his first weekly meeting as curator, he showed how glass bubbles containing partial vacuums "broke with a brisk noise." He also promised an experiment "about the tenacity of air."[19] The following week, he reported on his attempts to weigh air, and soon after that he performed an experiment using more glass balls, some unsealed and some evacuated and sealed. He conducted experiments suggested by other Fellows as well as his own. At one meeting, he was to prepare two experiments for the following Wednesday, and in the meantime to climb to the top of Westminster Abbey and weigh some items there and on the ground, to see if there was any difference. All of this he duly delivered. At a meeting on January 21, 1662, he proposed "to bring in at the next meeting the following experiments: 1. Of the living of insects in condensed air. 2. Of the force of falling bodies. 3. Of respiration. 4. Of the different refractions in cold and warm water."[20] He pursued all of them, while at the same time coming up with forty-eight queries to be sent to a correspondent living in Iceland. Would quicksilver congeal in the cold? What kind of substances were cast out of the burning mountain? How did whales breathe? Were there spirits, and if so what shape were they, and what did they say or do?

Hooke's output was astonishing, especially in these early years as curator of experiments. And until the 1670s, when he began to suspect that his efforts weren't appreciated enough, his commitment to the Royal Society was steadfast. In *Micrographia*, his great 1664 work on experimental microscopy, he went out of his way to proclaim his association with the Society, describing himself on the title page as a Fellow. In his dedication to Charles II he declared that several other Fellows were busy preparing books on the improvement of manufactures and agriculture as well as on navigation and the increase of commerce. He even produced a second dedication, to the Society itself, in which he offered "these my poor labours to this most illustrious assembly."[21] It was noted that after the Great Fire of London in September 1666, when various designs for the rebuilding of the city were being produced, Hooke presented his scheme to the Society for its approval, while Christopher Wren took his own proposal straight to the king. When the Society's secretary, Henry Oldenburg, finally got to see Wren's scheme, he noted with chagrin that "such a model, contrived by him, and reviewed and approved by the Royal Society, or a Committee thereof, before it had come to the view of his Majesty, would have given the Society a name, and made it popular, and availed not a little to silence those, who ask continually, What have they done?"[22]*

It was Hooke who provided the impetus for experimentation during the Society's first decade. In 1664, about seventy-five experiments were performed at meetings, with around the same number being conducted elsewhere and reported on. Hooke wasn't solely responsible; the medical Fellows were also particularly active, with research into the effects of poison, the physiology of respiration, and, most famously, the transfusion of blood. On November 22, 1667, Richard Lower and Edmund King reported to the Society that a Cambridge graduate named Arthur Coga was prepared "to suffer the experiment of transfusion to

---

*Like all the others, Wren's scheme was finally rejected in favor of a piecemeal reconstruction of the old city. More than a century later, elements of his plan surfaced in the design by Thomas Jefferson and Pierre Charlies L'Enfant for a new city on the bank of the Potomac River, Washington, DC.

be tried upon himself for a guinea."[23] Two days later, they carried out the procedure at Arundel House in front of an audience of "many considerable and intelligent persons."[24] Having opened the carotid artery in a young sheep and a vein in Coga's arm, they reported, "we planted our silver-pipe into the said incision, and inserted quills between the two pipes already advanced in the two subjects, to convey the arterial blood from the sheep into the vein of the man."[25] By some miracle Coga survived. He presented himself before the Society on November 28 to give an account "of what he had observed in himself since he underwent the said experiment."[26] He also told members that he would be happy to undergo the experiment a second time, presumably on the production of another guinea.

These medical researches were illuminating, although often distressing: kittens and puppies were regularly killed or had organs removed at Society meetings in all kinds of horrible ways. In 1664, Hooke vivisected a dog: using a pair of bellows and a pipe inserted into its trachea, he "was able to preserve [it] alive as long as I could desire, after I had wholly opened the thorax, and cut off all the ribs, and opened the belly."[27] The suffering he caused the creature made him swear never to repeat the experiment, although in fact he did three years later, after medical members were unable to perform it successfully.

The flow of experiments slowed in 1665, when the plague caused many of the Society's members to decamp to the relative safety of Oxford; and then again after the Great Fire of 1666, when they were temporarily evicted from their premises in Gresham College. Although they were rehoused in Arundel House on the Strand, the facilities there were no match for their previous quarters. Experimentation never again reached the heights it had in Hooke's first years as curator. Although he remained active in the Society, his interests now lay elsewhere—he was assisting Wren with the rebuilding of the city's churches and pursuing his own interests in mechanics. In some years, only a dozen experiments were performed at meetings, and by the 1680s the council was looking for a replacement curator. In 1687, Hooke proposed that if his salary were raised to one hundred pounds a year, he would agree to "produce one or two experiments and a discourse at every meeting." But the proposition was turned down, and the Society did without a curator for the rest of the century.[28] It was steadily moving away from its initial incarnation as a practical forum for experiments and toward becoming a talking shop, where papers were presented but science was not.

4

PHILOSOPHICAL
TRANSACTIONS

*"Desires after solid and useful knowledge
may be further entertained"*

On Monday, March 6, 1665, a new publication appeared in London. Dedicated to the Royal Society and claiming to give "some accompt of the present undertaking, studies, and labours of the ingenious in many considerable parts of the world," *Philosophical Transactions* contained in its sixteen pages an eclectic mixture of articles, observations, and extracts from letters.[1] There was national and international news from the world of astronomy, including a long account from

France of astronomer Adrien Auzout's claim to have predicted the motion of the comet that appeared over Europe at the end of 1664, and a short notice of Robert Hooke's observations in May 1664 of a spot "in the biggest of the 3 obscurer *Belts* of *Jupiter*." There was a preview of—actually more of an advertisement for—Robert Boyle's *Experimental History of Cold*; without a hint of irony, the announcement declared that the publication had been delayed by "the extremity of the late frost," which had stopped the press. This was followed by an account sent to Boyle from Hampshire of "a very odd monstrous calf," a deformed fetus that had been discovered when its mother was slaughtered. And there was a description of whale-hunting in the Bermudas, an account of the potential of pendulum watches for calculating longitude at sea, and an obituary of the great mathematician Pierre de Fermat, "one of the most excellent men of this age," who had died just two months earlier.[2]

Philosophical Transactions was not only the first scientific journal in the world, but also became the longest running and arguably the most important. "If all the books in the world, except the *Philosophical Transactions*, were destroyed," wrote the great English biologist T. H. Huxley in 1870, "it is safe to say that the foundations of physical science would remain unshaken."[3] It was the brainchild of the Royal Society's secretary Henry

Oldenburg, who had been tutor to Robert Boyle's nephew Richard Jones. Oldenburg, born in Bremen, Germany, in about 1619, had spent the mid-1650s in England as Bremen's envoy to the Cromwellian court before receiving the tutoring post. He had accompanied Jones to Oxford in 1656 and then, having made the acquaintance of John Wilkins, and, through Wilkins, the group of experimental philosophers then gathering in Oxford and London, including Robert Hooke, Christopher Wren, Seth Ward, and John Wallis, took his pupil on an extended grand tour, returning to England in time for the Restoration of Charles II in May 1660. He remained close to Robert Boyle, seeing many of Boyle's books through the press and translating some of them into Latin while at the same time establishing correspondences with important European figures, including the philosopher Baruch Spinoza and the astronomer and mathematician Christiaan Huygens.

When the Society received its first royal charter in 1662, Oldenburg was named as a member of the council. He was appointed as one of the two secretaries (John Wilkins was the other). The exchange of knowledge was at the heart of the Society's work, and the charter explicitly licensed it "to enjoy mutual intelligence and knowledge with all manner of strangers and foreigners."[4] Oldenburg, who was more assiduous in his secretarial duties than Wilkins (the latter acted more as a vice-president than a secretary), operated as a clearinghouse for information and queries that came in from a network of correspondents all over the Western world. When he heard in 1664 that the French were talking of publishing "a journal of all what passeth in Europe in matters of knowledge both philosophical and political," it was the spur he needed to do something of the sort himself.[5]

The result was *Philosophical Transactions*, which was to appear on "the first Monday of every month," although in practice its appearances were sporadic. Licensed by the Royal Society and containing notices of work by many of the Fellows, the journal was nevertheless a private venture, and Oldenburg was at pains to point out that the *Transactions* "are edited and published by me alone."[6] He hoped to make £150 a year out of the project, although he rarely made more than a third of that amount: the printers sometimes refused to print the journal in the quantities they promised, convinced that they would never sell, and he often ended up giving away copies to foreign correspondents in return for their own publications.

The *Transactions* may not have filled Oldenburg's pockets quite as he had wished, but they changed the world. Yes, his motives were mixed. As well as seeing a profit in the venture, he wanted to advertise the good work being done by the Royal Society and its Fellows.

He was well aware that when it came to the furious and frequent arguments over priority besetting the scientific world, it was useful to be able to point to a dated, published source that could prove beyond doubt the point of first contact with a particular discovery. But the *Transactions* did much more than this: they established a public forum for the sharing of ideas, and in so doing, they created an international culture of knowledge exchange.

In the introduction to that first issue, the secretary set out his hopes for the new publication, which would not only describe current research and new discoveries in natural philosophy but also spur on the efforts of others:

> To the end, that such productions being clearly and truly communicated, desires after solid and useful knowledge may be further entertained, ingenious endeavours and undertakings cherished, and those, addicted to and conversant in such matters, may be invited and encouraged to search, try, and find out new things, impart their knowledge to one another, and contribute what they can to the grand design of improving natural knowledge, and perfecting all philosophical arts, and sciences. All for the glory of God, the honour and advantage of these kingdoms, and the universal good of mankind.[7]

The 142 issues of the *Transactions*, published, and partly written, by Henry Oldenburg between 1665 and his death in 1677, contain the occasional wonder. Issue 89, for example, included a paper by Boyle on luminescence in meat, a subject that would make the Society the butt of jokes in years to come. Issue 44 offered an account of a postmortem carried out by William Harvey on a man who had died at the age of 152 years and 9 months. Harvey attributed the man's death to a change of air, since he had recently been brought from the countryside to live in London, and a change of diet, because he had taken to eating rich foods. He also noted that his genitals were unimpaired, "serving not a little to confirm the report of his having undergone public censures for his incontinency."[8] But these oddities should not distract us from the fact that the majority of the papers appearing in these first issues of the *Transactions* were good, groundbreaking science. Giovanni Cassini's discovery of Saturn's moons was reported here. In 1672, Issue 80 contained "a letter of Mr Isaac Newton, Professor of the Mathematics in the University of Cambridge, containing his new theory about light and colour." Issue 81 included Newton's description of the reflecting telescope he had invented.

For a time after Oldenburg's death in 1677, it seemed that the *Transactions* had died with him. Over the next six years, Robert Hooke sought to fill the gap with occasional volumes of *Philosophical Collections*, which contained book reviews and "an account of such physical, anatomical, chemical, mechanical, astronomical, optical, or other mathematical and philosophical experiments and observations as have lately come to the publishers' hands."[9] Like the *Transactions*, Hooke's *Collections* relied heavily on the work of Royal Society Fellows, although it remained an unofficial publication. Hooke promised not to make use of anything contained in the register books of the Society without leave of the council and the author. *Collections* also included foreign intelligence, including letters from the Dutch microscopist Antonie van Leeuwenhoek, and it had its fair share of curiosities. Volume 2, for example, published a report about the birth in Somerset of conjoined twins; and another from Nuremberg about a body that had been buried for forty-three years, but which, when it was exhumed, "was found almost wholly converted into hair."[10] A sample of this hair was shown at one of the Society's weekly meetings and placed in the repository. It stimulated one of the Fellows, the physician Edward Tyson, to submit some anatomical observations on hair, teeth, and bone, which were also published in the *Collections*.

Hooke couldn't afford to invest heavily in the *Collections*, and print runs were small. The last issue, which contained work by van Leeuwenhoek, Jacob Bernoulli, and Gottfried Leibniz, appeared in April 1682. The following year the *Transactions* resumed, under the joint editorship of Francis Aston and Robert Plot. Both men were secretaries of the Royal Society at the time, but it was still not an official organ of the Society, as the preface to the first volume of the new series made clear: "The writing of these transactions, is not to be looked upon as the business of the Royal Society."[11] However, the preface went on to emphasize that it was at the Society's behest that the *Transactions* were being revived; they were "a convenient register, for the bringing in, and preserving many experiments, which, not enough for a book, would else be lost."[12]

Philosophical Transactions has been a central feature of the Royal Society's publishing program ever since. Its content has changed according to the tastes of its editors, the availability of material, and advances in scientific research. In 1752, eighty-seven years after Oldenburg's first issue, the Royal Society finally took over official responsibility for the *Transactions*. Up until then the editorship and choice of papers to include had fallen on successive secretaries; the Society had steadfastly maintained that it was not in any way responsible for them, even though the rest of the world had always viewed the *Transactions* as an official Royal Society publication. But in January 1752, the physician and antiquary Cromwell Mortimer, who, as secretary, had edited the *Transactions* since 1730, died; and the following month the Earl of Macclesfield, who was on the council at the time, proposed that the Society take it in hand. A twenty-one-member Committee of Papers was set up to vet contributions and select those suitable for publication. This action marked rather more than an acknowledgment of the status quo: all papers read at the weekly meetings were now automatically considered for publication by the committee, so that the appearance of a paper in *Philosophical Transactions* was a seal of official approval. In fact, a paper was rarely read in its entirety, either before the Society or in committee; instead, judgment was usually based on an abstract of three hundred to five hundred words. Nor were the papers discussed or debated at gatherings of the full membership. All that members were required to do was listen, which, as we shall see, led to some mind-numbingly boring meetings.

In 1776 the journal was renamed, becoming *The Philosophical Transactions of the Royal Society of London.* By the later nineteenth century, it had become one of the world's leading scientific forums: Darwin and Huxley were published in it; Humphry Davy's description of his safety lamp for miners appeared there; so did Michael Faraday's experimental researches in electricity, Benjamin Franklin's account of his electrical kite experiment, and Charles Babbage's description of the forerunner of the computer, his "difference engine." Edward Jenner had two papers in the journal, although neither gave a hint of Jenner's future as a pioneer of inoculation against smallpox: one was about the migration of birds, the other an observation on the natural history of the cuckoo. In 1887, the demand for space was so great that the *Transactions* were split in two and named *Philosophical Transactions of the Royal Society of London A and B*, a distinction that became clearer to outsiders in 1896: "Series A, Containing Papers of a Mathematical or Physical Character," and "Series B, Containing Papers of a Biological Character." And *A* and *B* they remain, although *A* now includes the engineering sciences.

In the meantime, the publishing function of the Society grew. The first expansion took place in 1800, when anyone whose contribution had been accepted for the *Transactions* was required to submit an abstract, which appeared in a new publication called, with commendable clarity, *Abstracts of the Papers Printed in the Philosophical Transactions of the Royal Society of London*. This title later developed into *Abstracts of the Papers Communicated to the Royal Society of London* and then *Proceedings of the Royal Society of London*, which were also split into *A* and *B* along the same lines as the *Transactions*. In 1938 a third periodical appeared. *Notes and Records of the Royal Society of London* began as a house journal, with news of various dinners and other functions and historical articles of interest to the Fellows. It is now devoted exclusively to the history of science and is an essential source for anyone researching in the field.

In the twenty-first century, Royal Society Publishing is home to a thriving array of titles. In addition to *Transactions A* and *B*, *Proceedings A* and *B*, and *Notes and Records*, there is *Biographical Memoirs of Fellows* (actually a collection of obituaries of the recently deceased), which is published annually. *Biology Letters* publishes short articles on the biological sciences; *Interface* looks at the interface between the physical and life sciences; and its offspring, *Interface Focus*, consists of specially themed issues. *Royal Society Open Science* welcomes "high-quality science including articles which may usually be difficult to publish elsewhere, for example, those that include negative findings."[13] *Open Biology* is an online journal publishing high-impact biology at the molecular and cellular level. All the Society's journals, apart from *Notes and Records* and *Biographical Memoirs*, now operate under a continuous publication model in which the online version is the authoritative record. And all still fulfill Henry Oldenburg's claim to give "some accompt of the present undertaking, studies, and labours of the ingenious in many considerable parts of the world."

5

REPOSITORIES AND LABORATORIES

"A cabinet of curiosities, very full, but well kept up"

Solomon's House didn't remain at Gresham College for long. The Great Fire that swept through London in September 1666 left Gresham, along with most of the streets in the northeast corner of the city, unscathed. But everything else within the city walls was destroyed or badly damaged: 13,200 houses, 86 churches, 44 of the 51 livery company halls, and the Custom House, the Guildhall, the General Letter Office, and the Royal Exchange.

The Society's regular meeting rooms were commandeered by the city authorities, and in January 1667, the Fellows, after retreating to the lodgings of Walter Pope, who had succeeded Wren as Gresham professor of astronomy, took up an offer from Henry Howard to move their weekly meetings to Arundel House, a sprawling medieval palace between the Strand and the Thames belonging to Howard's brother, the 5th Duke of Norfolk.

Howard offered them a site on the grounds that was thirty meters long by twelve meters wide (about one hundred by forty feet). For some time, several of the Fellows had discussed the Society having its own purpose-built premises, an institute or college that would serve, as Thomas Sprat wrote, "for their meetings, their laboratories, their repository, their library, and the lodgings for their curators."[1] Christopher Wren produced an ambitious scheme for a five-story building containing everything a group of experimental philosophers could desire: along with a galleried two-story meeting room ("very useful in case of solemnities"), it would have a laboratory and a library, workshops, a smithy, a long attic gallery where lenses could be tried, and a flat roof for testing telescopes.[2] A central cupola could double as an astronomical observatory and an anatomy theater. Gaps could be left in the floors from attic to cellar for experiments with barometric pressure, the velocity of falling bodies, and the vibration of pendulums.

Wren's college came to nothing. It would have cost more than £4,000 at a time when all attention and resources were focused on rebuilding London after the Great Fire. A stripped-down scheme offered by Robert Hooke was adopted instead, but that, too, fizzled out for lack of funds, and in 1673 the Society moved back into Gresham College.

It remained there for another thirty-seven years, until 1710. As early as March 24, 1703, just three weeks after Hooke's death, the Gresham trustees made it clear that they would like the Society to move out. The college was anxious about the state of the building and wanted to rebuild it. The Fellows, however, stalled for the next seven years, lamenting that "the very embryo of the Society had been formed in Gresham College." They tried to find alternative premises, but without much success.[3] By September 1710, the Gresham trustees were becoming quite insistent that they really must leave. The president of the Society at the time, Sir Isaac Newton, summoned the council and informed them that a house belonging to "the late Dr Brown," in Crane Court, just off Fleet Street, was up for sale: "And being in the middle of the town, and out of noise, [it] might be a proper place to be purchased by the Society for their meetings."[4] The upshot was that in October the Society purchased Crane Court for £1,450 in spite of opposition from members who thought it was rather poky in comparison with their rooms at Gresham. The first meeting was held there on November 8, 1710. Soon afterward, they had a galleried repository built in the garden, quite possibly to the designs of the seventy-eight-year-old Sir Christopher Wren, which would have made it his last building. "That noble body have removed into Two Crane Court in Fleet Street," reported John Macky in 1722, "where they have purchased a very handsome house, and built a repository for their curiosities, in a little paved court behind."[5] In order to lend what was a fairly small house in a narrow court some of the gravitas lost in the move from Gresham, Newton ordered that the Society's porter be gowned and provided with a staff, which would be surmounted with the Society's arms in silver; on meeting nights, a lamp was hung over the entrance into the court from Fleet Street.

The Royal Society remained at Crane Court for seventy years. But as members of an established national institution, as eager as ever for their status to be recognized by others, the Fellows periodically expressed a desire for premises that were rather grander than a house tucked away down an alley, even if there was a gowned porter brandishing a staff at the door. The opportunity came in 1775, when Parliament gave Buckingham House— later to become Buckingham Palace—to George III's queen, Charlotte, in return for her giving up her rights to Somerset House on the Strand, the remains of a vast Tudor palace that was traditionally the residence of queen consorts (even though no queen had used it for almost a century). In the same year, an act was passed for demolishing Somerset House and replacing it with a national building to house a range of public offices, from the Navy Office and the Tax Office to the King's Bargemaster's House. Sir William Chambers, surveyor-general of the king's works, was given the job of designing the new building, and work had scarcely started before the government let it be known that it intended to offer rooms there to the Royal Society.

The rooms proposed weren't good enough, in the opinion of the Fellows. "We would wish to have it considered that the allotment of public apartments to the Royal Society," ran a letter signed by the president and the entire council, "will be understood by all Europe, as meant to confer on them an external splendour."[6] Yet there was hardly enough space for the Society's collection of books. And there was no room at all for its Repository.

The Society's museum, always known as the Repository—"a general collection of all the effects of arts, and the common, or monstrous works of nature"—had been a great source of pride in the early days.[7] Members donated to it all kinds of curious artifacts: a bird of paradise, an ostrich egg, a piece of petrified wood.

Calls occasionally went out for specific kinds of items: when Sir Robert Moray promised to send in a piece of copper ore from Sweden, for example, members who had access to similar rock types were asked "to bring in ores of several kinds, to be put into their repository."[8] In 1664, Christopher Merrett and Walter Charleton, both medical men, were asked to make a list of exotic and domestic animals suitable for inclusion, along with instructions on how to preserve their skins.

The scale, if not the nature, of the Repository changed dramatically in 1666, when a donation of one hundred pounds from the Society's wealthy treasurer, Daniel Colwall, enabled it to buy one of the most famous collections of curiosities in England. Brought together "with great industry, cost, and thirty years travel in foreign countries," Robert Hubert's collection of natural rarities was one of London's sights.[9] Every afternoon, his house near the west end of St. Paul's Cathedral was open to visitors on payment of a fee. Private tours were offered in three or four different languages, and great emphasis was placed on the important royal and aristocratic characters who had seen the museum at one time or another—emperors, empresses, kings, queens, and princes. There was a mummy "adorned with hieroglyphics [and] taken out of one of the Egyptian pyramids";[10] the thigh-bone of a Syrian giant; a two-headed calf and the paw of a polar bear; an extraordinary array of exotic birds, including "a marvellous great head and bill of a bird yet unknown";[11] fish and sea shells; serpents and lizards; oddly shaped vegetables, including "two very perfect mandrake roots";[12] minerals and gemstones; and "things of strange operation," including a mineral substance that, according to Hubert's comprehensive catalog, "being put into a glass of wine, makes infinite bubbles like atoms that rise in the middle of the wine, to the delighting of the beholders."[13]

The motives of the Royal Society's members in acquiring Hubert's collection were mixed. Some reveled in the sheer joy of owning such a ready-made cabinet of curiosities, although, from early on, the Society was keen to make it more respectable, describing Daniel Colwall as the founder of the museum and erasing the showman Robert Hubert from the narrative. Others saw this dramatic expansion of the Repository (Hubert's catalog listed around one thousand objects) as an opportunity for scholarship, agreeing with Hooke that such things were not "for divertisement, and wonder, and gazing . . . but for the most serious and diligent study of the most able proficient in natural philosophy."[14] But everyone saw that the Repository was a means of enhancing the young Royal Society's status. An institutional museum had continuity, unlike private collections, which were liable to be dispersed upon the deaths of their owners. This collection offered academic respectability, and it was a means of attracting donors.

The high point for the Society's Repository came in 1681 with the publication of secretary Nehemiah Grew's *Musaeum regalis societatis* (Museum of the Royal Society), "a catalogue and description of the natural and artificial rarities belonging to the Royal Society and preserved at Gresham College."[15] Running to 388 pages, Grew's catalog included many of Hubert's specimens (though not all—some had been lost or had decayed beyond rescue), as well as others donated by Fellows. Grew was careful to name the donor wherever possible. A waywiser was given by John Wilkins. A model of a roof that involved some cutting-edge structural engineering was provided by John Wallis. It was this design, with its complicated geometrical pattern, that Wren used in the ceiling of the Sheldonian Theatre in Oxford. Sir Joseph Williamson, the third president of the Society, contributed the horn of a sea unicorn. It was 2.5 meters (8.2 feet) long and "very beautiful in length, straightness, whiteness, and its spiral furrows."[16]

Grew provided meticulous descriptions of each item, referencing, and sometimes disputing, earlier authorities. He also corrected Hubert, identifying the thigh-bone of a Syrian giant as belonging to an elephant. Hubert's mummy had disappeared, either sold off separately or lost in transit. But the Society had one of its own, presented by Henry Howard in 1667, and Grew described it in great detail. The hieroglyphics were figures of men, women, and birds, in gold, yellow, red, and blue—and not very well executed, he noted. The mummy itself was 1.6 meters (5.2 feet) long, and the inner winding sheet (of three) was stained with "a blackish and gummous substance." From this, Grew deduced that "the way of embalming amongst the *Aegyptians*, was by boiling the body (in a long cauldron like a fish-kettle) in some kind of liquid balsam; so long, till the aqueous parts of the flesh being evaporated, the oily and gummous parts of the balsam did by degrees soak into it, and intimately incorporate therewith."[17] It was much like preserving pears in sugar, he suggested.

The Royal Society's Repository, accessible to outsiders in a way that the Society's regular meetings were not, became something of a tourist attraction. Christiaan Huygens, who visited in 1689, talked of seeing "a cabinet of curiosities, very full, but well kept up";[18] and in 1708 the *New View of London* devoted twenty pages to descriptions of "the most remarkable rarities in the Repository at Gresham College," mostly taken from Grew's catalog.[19] The author of *British Curiosities in Nature and Art* (1713) was delighted with the "collection of wonderful curiosities" at Crane Court, singling out Howard's mummy, Wallis's geometrical design, and "a stone voided by the penis of a man at Exeter, 2 inches and a quarter in length of a pyramidal form."[20] Others were less enthusiastic. The satirist Ned Ward, writing in 1703, dismissed the Repository as "rusty relics and philosophical toys."[21]

There is evidence that the early eighteenth century marked a downturn in the Repository's fortunes. In the summer of 1710, when Gresham was pushing the Royal Society to find new premises, the German tourist Zacharias Conrad von Uffenbach complained that the artifacts "were not only in no sort of order or tidiness but covered with dust, filth and coal-smoke and many of them utterly broken and ruined."[22] The move to Crane Court a few months later and the construction of a new building to house the Repository do not seem to have done much to improve matters: in 1729, a committee established to consider its condition found some of the specimens "decayed, and the rest of them on great disorder."[23] The following year a writer dismissed them as "extraordinary gugaws and insignificant rarities."[24]

Over the next fifty years, various attempts were made to put the Repository in order. But the task required resources that the Society didn't have and a commitment on the part of individual members that couldn't be sustained. The assemblage of objects hovered between a research collection and an old-fashioned cabinet of curiosities. Ironically, considering that the founders had believed that an institutional collection offered more stability and continuity than a private museum could, members came to see how hard it was for an institution like theirs to take continuing responsibility for a permanent project. In 1752, the Society's librarian and keeper of the Repository, Emanuel Mendes da Costa (later to achieve notoriety by embezzling £1,500 of the Fellows' subscriptions and spending five years in debtors' prison as a result), reported that "foreigners of curiosity, as well as our own peoples, often desired to see our museum, which had formerly a reputation both at home and abroad. [I] was ashamed to recite what a ruinous, forlorn condition it was now in, and prayed it would be amended."[25]

Ten years later a committee was again tasked with reporting on the state of the collection. It found that many of the animal and vegetable specimens were in an advanced state of decay. The animals, in particular, were in such a state of putrefaction, said the committee, that "the air in the room became intolerably foetid, and they were all sick."[26]

So when the crown offered the Royal Society premises in the new Somerset House, it was—for some members at least—a relief to find there was no space for the Repository. Fortunately, a new home for the collection soon became available. In 1753, the great collector

Sir Hans Sloane, who had been president of the Royal Society from 1727 to 1741, died, leaving his entire collection, some seventy-one thousand objects, to the nation and effectively creating the British Museum, which opened to the public in 1759. At some point in the late 1770s, while the move to Somerset House was being discussed, the Society's council decided to transfer the Repository to the British Museum. It did not put the decision to the members, presumably foreseeing that some would disagree.

When the Royal Society moved into Somerset House in 1780, the Repository did not move with it; the following year, the entire collection, which by now numbered somewhere in the region of six thousand objects, was handed over, lock, stock, and putrefying barrel, to the British Museum. A letter from the museum's trustees was read out at a meeting in November 1781, thanking the Society for its donation of "a very ample collection of natural productions."[27]

6

PERSONS OF
GREAT QUALITY

*"If I have seen further, it is by standing
on the shoulders of giants"*

IN ITS EARLIEST DAYS, THE ROYAL SOCIETY'S DIREC-
tor was chosen monthly, and the post tended to go
alternately to founders John Wilkins and Sir Robert
Moray. But when the first royal charter of 1662 stipulated that the Society should consist of "a President,
Council, and Fellows," it also named as its first president "our very well-beloved and trusty William, Viscount Brouncker, Chancellor to our very dear consort,
Queen Catherine."

Forty years old, Brouncker was the son of a minor Anglo-Irish courtier whose devotion to the Royalist cause had earned him an Irish peerage in September 1645, a few months before his death at Oxford. Charles II was repaying old debts when he made the younger Brouncker chancellor to Catherine of Braganza; his appointment as first president of the Royal Society may also have been a reflection of royal gratitude for the loyalty of Brouncker *père*. But Brouncker was also a qualified medical doctor—he had been awarded the degree of doctor of physic at Oxford in 1647, although there is no evidence that he had ever practiced—and he was certainly a talented mathematician. While others followed Charles II into exile, he had spent the Commonwealth years keeping his head down and quietly pursuing his mathematical studies. Like many other mathematicians, then and now, he took an active interest in musical theory, and he published an anonymous translation of Descartes's *Musicae Compendium* in 1653. In the later 1650s he collaborated with John Wallis, and it was perhaps through Wallis that he got to know other members of the Oxford group. At the Restoration he began to attend the Gresham lectures along with the returning Royalist exiles Sir Robert Moray and Alexander Bruce, which was how he came to be there at the foundation meeting on November 28, 1660. He was respected by the king and by his peers.

And he was no figurehead. He proposed experiments, conducted experiments of his own, and assisted Boyle and others in theirs. The Frenchman Samuel de Sorbière, who visited London in the summer of 1663 and was introduced to the Royal Society by Moray, described one of the Wednesday afternoon meetings at Gresham at which Brouncker presided. The meetings took place in a large wainscoted chamber, he said, where two tiers of bare wooden benches, one slightly higher than the other, were arranged before a table in front of the fire. Seven or eight chairs upholstered in green cloth were arranged around the table; they remained unoccupied on the occasions when de Sorbière was present, leading him to suggest they were reserved "for persons of great quality."[1] Brouncker, as president, sat at the table in the room's only armchair, with the secretary—presumably Henry Oldenburg—on his left, taking minutes. All the other members took their places on the forms "as they think fit, and without any ceremony; and if any one comes in after the Society is fixed, nobody stirs, but he takes a place presently where he can find it, that so no interruption may be given to him that speaks."[2] Fellows stood to address the chair bare-headed until Brouncker signaled for them to replace their hats, and everyone was civil, respectful, and polite. No one was ever interrupted, and if anyone dared whisper to his neighbor, "the least sign from the president causes a sudden stop, though they have not told their mind out."[3]

The charters stipulated that the presidency should run for one year, with elections taking place at the annual St. Andrew's Day meeting on November 30. But they did not set a maximum number of terms, and Brouncker, who was returned unopposed November after November, failed to notice that he had outstayed his welcome. By the mid-1670s, some of the Fellows clearly thought it was time for a change at the top: in 1677, a group of disgruntled Fellows engineered a coup, petitioning Sir Joseph Williamson, Charles II's secretary of state, to stand for the presidency. Brouncker hadn't shown his face at a council or at the weekly Gresham meetings for months. When he found out about Williamson, he saw that he was facing a contested election, one he was likely to lose, and stood down, angry and hurt. Williamson, who had been elected a Fellow in 1662, was returned unopposed, the first of many Royal Society presidents to be chosen more for their political influence than for their science. Sir Christopher Wren was made vice-president, often taking the chair during Williamson's three-year tenure when affairs of state meant he had to absent himself.

All political careers end in failure, as the saying goes. Williamson's was no exception. He fell spectacularly from grace in 1679 when he was caught up in the hysteria surrounding the Popish Plot, which falsely alleged that there was a Catholic conspiracy to assassinate Charles II. Williamson, briefly imprisoned in the Tower, lost his post as secretary of state, and when the time came to reelect him for a fourth term as president, in 1680, he let it be known that he wasn't going to stand. The Society cast around for a successor and approached Robert Boyle. Evelyn reckoned Boyle should have been president from the first, and said that now "neither his infirmity nor his modesty could any longer excuse him."[4]

Boyle accepted, but there was a snag. The anti-Catholic Test Act of 1673 stipulated that all officeholders had to swear allegiance to the monarch as head of the Church of England and repudiate the Catholic doctrine of transubstantiation. As a fervent Protestant, Boyle didn't have a problem with either of these ideas, but like many fervent Protestants of the time, he *did* have a problem with taking an oath. He took his cue from James 5:12: "Above all things, my brethren, swear not, neither by heaven, neither by the earth, neither by any other oath: but let your yea be yea; and your nay, nay; lest ye fall into condemnation." He agonized for several weeks and finally informed the Society that he had to decline the honor after all.

This left an embarrassing vacuum. After some hurried negotiations, it was filled by Sir Christopher Wren, who was sworn in as president at a council meeting on January 12, 1681. Wren remained at the helm for the next two years in spite of a demanding day job: he was in the middle of rebuilding St. Paul's Cathedral after the Great Fire, and in addition, he was leading the team that was building fifty-six new parish churches for London. He was also surveyor of the king's works, which made him the most senior architect in the kingdom, and from 1682 he was occupied with the creation of a vast new palace for Charles II at Winchester, on a scale to rival Versailles.* Even so, he played an active part in the Society's proceedings, giving papers on topics ranging from Chinese medicine to the natives of Hudson's Bay, who he claimed lived to be 140 years old—"without the use of spectacles."[5] He also established anatomical, agricultural, and cosmographical committees "to register all things, that should be remarkable." Moreover, he got to grips with the Royal Society's finances, which were in a poor state, largely because so many members were neglecting to pay their weekly subscription of one shilling.

Nonpayment of dues had been a problem since the beginning, with the Society caught between the need

---

*The King's House in Winchester was built between 1683 and 1685, but the project overall was abandoned after the death of Charles II. The building was used as a prisoner-of-war camp and then as a military barracks before being destroyed by fire in 1894.

to attract and keep distinguished scientists and influential courtiers, on the one hand, and the reluctance, on the other, of many of those scientists and courtiers to pay for the privilege of membership. Determined to grasp the nettle, Wren launched a campaign. As a first step, Fellows called on prominent defaulters and urged them to cough up their arrears. All kinds of excuses came back. Sir William Petty said his wife would pay. Ralph Bathurst, president of Trinity College and dean of Wells Cathedral, sent ten pounds and a pledge to leave the Society something in his will. The septuagenarian poet Edmund Waller, who hadn't paid a penny since 1663, sent a sad little message pleading his losses in the Civil War and the expenses of having so many children (he had fourteen). The courtier Sir Nicholas Steward, when told he owed eleven years' subscription money, insisted the Society had got its sums wrong.

Wren's reaction was to suggest the Society remove the names of everyone who was in arrears from the published list of members. This was too drastic a step: it would have meant the effective expulsion of around half the membership, including some powerful friends at court. The council therefore watered down the proposal, and used it instead to address a different problem. They decided to target a selection of twenty-three members, consisting chiefly of individuals whose departure would not be damaging to the Society either scientifically or politically. A number of men without much interest or expertise in the new philosophy had been joining the Society, almost as one might join a gentlemen's club. A new statute tried to counteract this tendency by ordering the council to examine prospective candidates for fellowship, to see "whether the person is known to be so qualified, as in probability to be useful to the society."[6] The new statute failed to solve the financial problem, however. For the next two hundred years the argument over the admission of high-status dilettanti would surface again and again.

Wren was the last leading player in the scientific revolution to occupy the post of president for the rest of the seventeenth century. Between 1682, when he stood down, and 1703, there were eight presidents, all of whom were politicians, lawyers, or administrators. They were prominent enough at court or in Parliament, but by no means professional scientists or academics. The Society began to drift and to lose focus.

This tendency was reversed, however, in 1703, when the 1st Baron Somers, a prominent Whig lawyer and past lord high chancellor of England, stepped down from his five-year stint as president. His place was taken by Sir Isaac Newton, the sixty-year-old master of the Royal Mint, an ex-Cambridge academic with an international reputation as a physicist and mathematician— and, in the words of a twentieth-century chronicler of the Royal Society, "the greatest name in the history of the Society."[7]

Brilliant, secretive, sensitive to any criticism, Newton had been a member of the Royal Society since January 1672, when he was elected after sending in a paper describing his reflecting telescope, which caused a sensation in the scientific community. Unfortunately, he had soon earned the enmity of Robert Hooke, who was always ready to patronize a fellow scientist—and to attack one when he felt his own area of expertise was being encroached upon. When Newton's "hypothesis explaining the properties of light" was read to the Society in 1675, Hooke stood up and said there was nothing new in it—"the main part of it was contained in his [i.e., Hooke's] *Micrographia*, which Mr. Newton had only carried farther in some particulars."[8] Newton was furious, and the incident was the start of a lifelong feud between the two men.

The first edition of Newton's *Philosophiæ Naturalis Principia Mathematica* (*Mathematical Principles of Natural Philosophy*), which set out his laws of motion and his law of universal gravitation, appeared under the Royal Society's aegis in 1687, although the astronomer Edmond Halley undertook the financing of the publication. Hooke claimed the *Principia Mathematica* plagiarized his own work, and the relationship between the two men remained difficult. Newton's famous remark,

"If I have seen further, it is by standing on the shoulders of giants," was addressed to Hooke—who immediately took umbrage, imagining it to be a slur on his small stature. It was only after Hooke's death in March 1703 that Newton was prepared to play a more active role in the affairs of the Society. Once he was elected president that November, he held the post until his death in 1727.

As one of the most prestigious experimental philosophers in the world, Newton enhanced the reputation of the Royal Society at a time when it was losing some of its original impetus. Of the original twelve men present at that first meeting in Rooke's rooms at Gresham College in 1660, only Christopher Wren and Abraham Hill were still living by the end of 1703. Boyle and other early supporters of the Society, including Henry Oldenburg and Samuel Pepys, were gone. Thomas Sprat, author of *The History of the Royal-Society of London*, was nearly seventy; John Evelyn was over eighty. Newton brought new life to the Society—not only through his scientific achievement, but by creating a culture of commitment and engagement, without which any collective institution is bound to fail. Considered by some to be imperious and arrogant, he found that the weekly meetings clashed with his work at the mint, and moved them from Wednesdays to Thursdays. But he regularly attended them, as well as the council meetings that preceded them. Out of

175 council meetings held between his election and the end of 1726, he was present at 161: three of his recent predecessors hadn't shown their faces at a single one.

There were nine presidents of the Royal Society between Newton's death in 1727 and the end of the eighteenth century, and they were a mixed bunch. Some achieved great things, both for the Society and for society at large. Sir Hans Sloane, for example, who succeeded Newton and held the presidency until 1741, was a capable administrator who actually managed to gather in some of the outstanding arrears owed by the Fellows—no mean task. He was also an avid collector in his own right, spending some £50,000 in buying "the rarities which every country produced." According to Thomas Birch, the eighteenth-century chronicler of the Royal Society, "his constant endeavour was to employ them to the best purposes, by making himself acquainted, as far as possible, with the properties, qualities, and uses, either in food, medicine, or manufacture of every plant, mineral, or animal, that came into his possession."[9] He left this collection, along with his manuscripts and books, first to George II and then, if he didn't want it, to the Royal Society. This was on the understanding that his two daughters should receive £20,000 from whomever acquired it. The king was not keen, and the Society couldn't afford it; Sloane's trustees decided to offer it to the nation instead, and it formed part of the founding collection of the British Museum, established in 1753.

Not all the eighteenth-century presidents are remembered for their leadership or their scientific achievements. The physician Sir John Pringle, president from 1772 to 1778, was best known for his habit of dozing off during meetings, according to an anonymous poet of the time:

> *If ere he chance to wake in Newton's chair*
> *He wonders how the devil he got there.*[10]

If the beginning of the century was dominated by Newton, the end of it was the preserve of the botanist Sir Joseph Banks, the longest-serving president in the Society's history, who was elected in 1778 and remained in office until his death in 1820.

Banks, born in 1743, had a passion for science and the wealth to indulge it. He was elected a Fellow in 1766 at the early age of twenty-three. He went on to fund himself as naturalist on several important expeditions, notably Captain James Cook's 1768 voyage in the *Endeavour* to the South Pacific to observe the transit of Venus and search for *Terra Australis*, a project supported by the Royal Society. He joined the council in 1774, and after his election as president four years later made an immediate impact, confirming the Society's role as national adviser to the government on scientific matters. He served as an active ex officio member of various governmental boards, including the Royal Observatory, the Board of Longitude, and the Board of Agriculture. A personal friendship with George III did no harm to his cause, either. His house at Soho Square became something of a scientific salon. Every Thursday morning he held a breakfast for guests with scientific interests, and every Sunday evening he held a *conversazione*, bringing together leading figures from the worlds of science and politics.

Behind this sociable front, though, Banks was autocratic and controlling, and fiercely protective of the Society's privileged position in British scientific life and of his own position as president. On one occasion, when the council was moving to replace him, he responded by dismissing the council instead. He managed elections so adroitly that admission to the Society depended on his good opinion. Those men he liked got in, irrespective of their scientific standing.

It was during Banks's term of office that a number of rival institutions were established: the Linnaean Society in 1788, the Royal Institution in 1799, the Geological Society in 1807, and the Royal Astronomical Society in 1820. They all had an impact on the Royal

Society, not least because research papers that had previously come to the Society for publication were now going elsewhere. By 1830, for example, it was being claimed that no botanical papers had been received for years, because they all went to the Linnaean Society instead. Sometimes Banks opposed these new societies; sometimes he supported them. Sir Henry Lyons, writing in the 1940s, believed that "so long as the new bodies were prepared to be definitely subordinate to the Royal Society and to be in effect controlled or administered by its Council he had no objection to them; but if they demanded an independent existence and complete freedom in the management of their affairs he opposed them."[11]

Difficult though he could be, under Banks's long presidency the Royal Society was woven ever more deeply into the fabric of British scientific life. But that achievement in itself brought criticism, as well as calls for the Society to return to its roots as a forum for the theory and practice of experimental philosophy.

7

STICKS AND STONES

"What have the Society done?"

ONE OF THE OVERRIDING CONCERNS OF THE EARLY Royal Society was a quest for recognition. Fellows who published findings in various scientific fields were urged to let their readers know they belonged to the Society: a move that validated both author and institution. Robert Boyle's *Some Considerations Touching the Usefulness of Experimental Natural Philosophy* (1664) declared on its title page that its author was "the Honourable Robert Boyle Esq; Fellow of the Royal Society." Hooke's *Micrographia*, published the following year, did the same. John Evelyn's important 1664 treatise on forestry, *Sylva*, not only informed its readers that it was first delivered to the Royal Society ("that illustrious assembly"), but also said it was published by the Society's "express order." And President Brouncker's official imprimatur was prominently displayed opposite the title page.

Public approval of the Society and its aims had been expected to follow naturally from the royal charters of 1662 and 1663. The public, however, did not comply. In his preface to the fourth edition of *Sylva* (now spelled *Silva*), which appeared in 1706, the year of his death, Evelyn launched into an angry defense of the Society's record, inveighing against the way in which people had been "led away and perverted by the noise of a few ignorant and comical buffoons . . . [who] are with an insolence suitable to their understanding, still crying out, and asking, What have the Society done?"[1]

His answer, after noting that any attack on the Society was an attack on the "honour of our royal founder"—an automatic response by Fellows to any criticism—was to say that the search for truth takes time.[2] He cited the many trials and experiments, the publications, and the improvements in astronomy, optics, botany, and other fields. "It is an evil spirit, and an evil age," he wrote, "which having sadly debauched the minds of men, seeks with industry to blast and undermine all attempts and endeavours that signify to the illustration of truth."[3]

An early attempt to answer that critical question, "What have the Society done?" was Thomas Sprat's *History of the Royal-Society of London*. Sprat, a Wadham graduate and a protégé of John Wilkins, was asked in 1663 to produce a statement of the Society's aims and objectives for publication. He had more or less finished it by November 1664, but Henry Oldenburg and the council, worried that his text didn't dwell at sufficient length on the Society's achievements, set up a committee to select suitably impressive research papers for inclusion. Publication was also delayed by the twin scourges of plague and fire that disrupted London life in 1665 and 1666, and the *History* didn't appear until 1667.

When it did, Sprat came out fighting, attacking the Society's detractors and asserting that "a higher degree of reputation is due to discoverers, than to the teachers of speculative doctrines."[4] The *History* is in three sections: an overview of the course of natural philosophy since ancient times, a narrative of the Society's origins, and a defense of experimentalism and the practical value of the new philosophy. Members' contributions were mostly technological, emphasizing the value of knowledge turned to practical ends. There was a scheme by Hooke for recording the weather, an account by Brouncker of experiments he had carried out on the recoil of guns, and another from the chemist

Thomas Henshaw on the making of saltpeter. Sprat also produced a long list of instruments that members had invented, from quadrants and pendulum clocks to "a new sort of spectacles, whereby a diver may see any thing distinctly under water."[5] He closed by saying that "while the Old could only bestow on us some barren terms and notions, the New shall impart to us the uses of all the creatures, and shall enrich us with all the benefits of fruitfulness and plenty."[6]

Sprat's robust defense of the Society's achievements provoked yet more attacks. Although he was careful to suggest elsewhere in the *History* that experimental philosophers didn't want to discard the past entirely, the rejection of ancient authority smacked of subversion. Moreover, the mechanistic approach that Sprat and other members of the Society promoted seemed to some to leave no room for God or the old religious certainties. Robert South, the University of Oxford's public orator, made a swingeing assault on the new science and, by implication, the Royal Society, in a sermon delivered at Westminster Abbey in 1667: "That profane, atheistical, epicurean rabble, whom the whole nation so rings of, and who have lived so much to the defiance of God, the dishonour of mankind, and the disgrace of the age which they are cast upon, are not indeed (what they are pleased to think and vote themselves) the wisest men in the world."[7]

By "magisterially censuring the wisdom of all antiq-uity . . . and, as it were, new modelling the whole world," by "forming into a kind of diabolical society, for the find-ing out new experiments in vice," they assured themselves of a place in hell.[8] South repeated his attack two years later in a long speech at the inauguration of Christopher Wren's Sheldonian Theatre in Oxford. Evelyn, who was present at the celebration, was furious at South's "mali-cious and indecent reflections on the Royal Society, as underminers of the University."[9] On the ceiling of the Sheldonian, a painted allegory of Truth putting Ignorance to flight might seem at first glance to be an endorsement of the Royal Society and the new philosophy, but the scene is dominated by unbending Theology, whose iron rod and tablets of stone take precedence over the celestial globes and telescopes around her.

The universities, bastions of Aristotelian authority, were among the Society's fiercest critics. The church wasn't far behind. There were still plenty of men in the second half of the seventeenth century who refused to believe the earth moved around the sun: in the 1670s, the Puritan divine John Owen said that heliocentrism was not compatible with holy scripture, a position maintained by the Roman Catholic Church until 1757. Others reckoned it made no difference: "The day begins no sooner, nor stays any longer with Ptolemy than with Copernicus."[10]

Sprat, who was an ordained priest, argued that Christianity and science were perfectly compatible and that a rational search for truth could only strengthen Anglicanism. In 1668, at the prompting of Henry Oldenburg, another clergyman, Joseph Glanvill, who became a Fellow in 1664, published *Plus Ultra: Or, the Progress and Advancement of Knowledge Since the Days of Aristotle*, in which he repeated Sprat's arguments in somewhat stronger terms. He also said the Society was willing to acknowledge the contributions made by "the learned ancients," but not willing "that they should have an absolute empire over the reasons of mankind."[11]

The Society could argue coherently enough against accusations that it fostered atheism, or ignored what was still relevant in the teaching of the ancients, or undermined the world order. Members found it much harder to defend themselves against a more insidious form of criticism—mockery. Sprat acknowledged the threat in his *History*: "I believe that New Philosophy need not . . . fear the pale, or the melancholy," he wrote, "as much as the humorous, and the merry. For they . . . may do it more injury than all the arguments of our severe and frowning and dogmatical adversaries."[12]

From the beginning, members had been particularly sensitive to mockery. Founding member William Petty never forgot an occasion in the Duke of York's chamber at Whitehall when Charles II roared with laughter at the men of Gresham "for spending time only in weighing of air, and doing nothing else."[13] Petty recalled the gibe ten years later, almost word for word, in a talk to the Society. But while their royal patron making a joke at their expense was hurtful enough, it paled in comparison with more public ridicule. In May 1676, the Duke's Company, a theater company with the patronage of the Duke of York, put on a new play by Thomas Shadwell at Dorset Gardens, Fleet Street. The eponymous hero of *The Virtuoso* was Sir Nicholas Gimcrack, "the finest speculative gentleman in the whole world." And Gimcrack's experiments had audiences in stitches. This virtuoso spent his time examining cheese-mites through his microscope and transfusing blood from a sheep to a madman: "The patient, from being maniacal or raging mad, became wholly ovine, or sheepish; he bleated perpetually, and chew'd the cud: he had wool growing on him in great quantities, and a Northamptonshire sheep's tail did soon emerge or arise from his anus."[14] Gimcrack read his Geneva Bible by the light given off from a putrescent leg of pork; he managed to cut an animal's windpipe and keep it alive by blowing air into its lungs with a pair of bellows; he was preparing a lunar atlas, and through the telescopes set up in his garden, he could see "all the mountainous parts, and valleys, and seas, and lakes in [the moon]; nay, the larger sort of animals, as elephants and camels;

but public buildings and ships very easily. They have great guns and have the use of gunpowder. At land they fight with elephants and castles. I have seen 'em."[15]

Ridiculous? Perhaps. But Gimcrack's experiments all had their source in the work of the Royal Society. The microscopic mites were a reference to Hooke's *Micrographia*. Thomas Coxe published his experiment in transfusing blood between dogs in the *Philosophical Transactions* for May 1667, and, as we have seen, Richard Lower and Edmund King transfused blood from a sheep to a man, Arthur Coga, later the same year. Boyle's "observations about shining flesh," caused by his servants' discovery of a luminescent neck of veal in his larder, appeared in the *Transactions* in December 1672. Hooke kept a dog alive with a pair of bellows and a pipe inserted into its trachea while he vivisected it in 1664. Christopher Wren made a model of the moon in the early 1660s, and Wilkins had published *The Discovery of a World in the Moon* back in 1638.

This kind of satire hurt more than arguments because it was so public. Hooke went to see *The Virtuoso* one Friday night in June 1676 and was horribly embarrassed. "People almost pointed," he wrote in his diary that night. "Damned dogs."[16] Shadwell's play was regularly revived over the next thirty years, and Gimcrack became a byword for the foolish natural philosopher. At the end of the seventeenth century, William Wotton, who became a Fellow in 1687, lamented the damage the play had done to "the men of Gresham." "Every man whom they call a virtuoso, must needs be a Sir Nicholas Gimcrack," he complained. "Nothing wounds more effectually than a jest; and when men once become ridiculous, their labours will be slighted, and they will find few imitators."[17]

The pain was all the greater because, although Shadwell's virtuoso could certainly be found among the members of the Society who lived far from London, who often corresponded with Oldenburg and regularly sent in unsavory specimens and accounts of monstrous births, he represented everything the Society fought against—the purely speculative part of learning, the idea of knowledge for its own sake without utility or use.

The misrepresentation of the Royal Society as a club for the Gimcracks of the world persisted. In 1704, Jonathan Swift's *Tale of a Tub* included a parody of modern science that used some rather odd examples drawn from *Philosophical Transactions*, and in 1726, when Swift was in the midst of a quarrel with Isaac Newton, he returned to his attack on science in *Gulliver's Travels*. Gulliver's third voyage takes him to the flying island of Laputa, the inhabitants of which are expert in mathematics, astronomy, and instrument-making. But they fail to make any practical use of their discoveries: their clothing doesn't fit because they take measurements with quadrant and compass instead of using a tape measure. Swift again ransacked the *Transactions* for examples of speculative science. The publication became something of a staple for satirists as the eighteenth century wore on. In a pamphlet of 1743 titled *Some Papers Proper to Be Read Before the R---l Society*, Henry Fielding poked fun at a paper the Society had published in January 1743 describing the Genevan naturalist Abraham Trembley's observations concerning the freshwater polyp *Chlorohydra viridissima*. He had written of the polyp's "surprising property, that being cut into several pieces, each piece becomes a perfect animal."[18] In Fielding's satire, however, the polyp becomes a guinea, and its curious reproductive habit a vehicle for an attack on misers and moneylenders.

One of the funniest portrayals of the Royal Society came from the pen of Sir John Hill, a physician and journalist with an interest in the natural sciences, whose failed attempts to join the Society led him to satirize its workings in a series of attacks. In 1750 he submitted to it a hoax treatise titled *Lucina sine concubitu* (*Pregnancy without intercourse*), in which he claimed to have made his chambermaid pregnant by dosing her with a preparation containing "animalcula," which he said he had discovered in the air. He also attacked the current president, Martin Folkes, writing that Folkes was a man of great ability in many branches of knowledge, but, "unluckily," none of them had "any connection with, or relation to, the business of the Royal Society."[19] He described how Folkes, "swollen to a double size with the honour of his office," had fallen asleep while presiding over a tedious meeting in Crane Court "from the natural effects of a full dinner, and a dull discourse."[20] When the meeting was over, most of the members decamped to a nearby coffeehouse to exchange information, but, Hill wrote, "We were hardly sat down, when one of the gentlemen began to tell us of a frightful monster, with wings and claws, voided by a lady, on taking a single dose of his worm-powder; a second, of a living wolf in one of his patient's breast; and a third, of a toad in a block of marble."[21]

Like the satirists who came before him, Hill had been consulting *Philosophical Transactions*. His *Review of the Works of the Royal Society of London*, which appeared in 1751, was a systematic dismantling of dozens of papers in the *Transactions*, from "a method to make fish shine" to "a new method of learning to sing" to an "an account of a merman."[22]

There is no evidence that these attacks deterred prospective Fellows. Nor did they do any lasting harm to the Royal Society as an institution. They hurt some members' self-esteem, but, if anything, they did a valuable service by highlighting a tendency, noticeable from the earliest days, for the Society to revel in the rare and the curious, searching for spectacle rather than serious science.

8

REFORM

*"The Society has, for years,
been managed by a party, or coterie"*

IN THE DECADE AFTER SIR JOSEPH BANKS'S DEATH
(and perhaps even in the decade before it), the Royal
Society's meetings at Somerset House fell into an un-
inspiring routine. According to the statutes, ordinary
meetings were held on Thursday evenings at eight
o'clock, and they lasted for about an hour "at the discretion of the president."[1] So each Thursday the Fellows—those who could be bothered to attend—filed
into the Society's meeting room at Somerset House and
watched as the president and the two secretaries took
their seats behind a large table on which Charles II's

huge silver-gilt mace was displayed. They waited while a list of the visitors who had asked to attend the meeting was read aloud, and while those visitors rushed in from a room next door and took their places on cross benches on either side of the hall.

Once everyone was settled, a secretary read the minutes of the previous meeting, "which consist in repeating, in fewer words, everything that was read by his colleague on a former evening," as one member put it.[2] At a nod from the president, the other secretary read out a list of candidates for election to Fellowship of the Society, and voting began. This involved the assistant secretary going around the room with the balloting-box for each Fellow to place a ball in the "yea" drawer or the "nay" drawer for the first on the list. While this was happening, an exchange of nods between secretary and president was the signal for the reading of a paper to begin, at which time the Fellows and visitors settled down to sleep. Their gentle snores were interrupted by the assistant secretary, who would wake up each "composed and quiescent Fellow for his vote." The reading of the paper was likewise interrupted, often midsentence, for the president to display the contents of the "yea" and "nay" drawers and announce the name of the successful candidate. If there were more candidates, the process, complete with interruptions, was repeated, until eventually the chiming of the Somerset House clock signaled an end to the meeting, "none of the Fellows present having . . . taken the least share in its proceedings."[3] There were no experiments, no discussions, no debates.

To its critics within the Society, this lack of activity was due to the habit of electing rich men to the Fellowship, in the expectation that although they might not know much about science, they could at least be expected to pay handsomely for the privilege. Successive presidents were quite open about this policy, sometimes carrying it to extraordinary lengths. When Sir Roderick Murchison was elected Fellow in 1826, for example, the president at the time, Sir Humphry Davy, explained carefully to him that his election owed nothing to his scientific work (he was a geologist) and everything to his personal wealth.

The year 1830 saw three very public demands for reform. The first came from Charles Babbage, Lucasian Professor of Mathematics at Cambridge, who is best known today as the inventor of the automatic calculating engine, an early form of the computer. Babbage devoted well over half of his 228-page book *Reflections on the Decline of Science in England and on Some of Its Causes* to the Royal Society and its faults. He launched a personal attack on the president, who by then was the applied mathematician and traditionalist Davies Gilbert—"Why Mr. Davies Gilbert became President of the Royal

Society I cannot precisely say"[4]—and cast doubt on the minute-taking abilities (and integrity) of the secretaries, citing instance after instance of incompetence and mismanagement. Babbage laid the blame for these abuses at the door of the president and the council, whom he characterized as a self-perpetuating elite:

> The Society has, for years, been managed by a *party*, or *coterie*. . . . The great object of this, as of all other parties, has been to maintain itself in power, and to divide, as far as it could, all the good things amongst its members. It has usually consisted of persons of very moderate talent, who have had the prudence, whenever they could, to associate with themselves other members of greater ability, provided these latter would not oppose the system.[5]

Babbage's solution was for the Society to pay more attention to regulating the numbers admitted as Fellows, and "to make it an object of ambition to men of science to be elected into it." He advocated distinguishing publicly between members who actually contributed papers to the *Philosophical Transactions* (in the late 1820s this amounted to 109 out of a total membership of 714), and those who didn't, a plan the council rejected on the grounds that those who didn't contribute wouldn't like it.

Reform

Reflections on the Decline of Science appeared in May 1830. Babbage presented it at a stormy meeting of the Society on May 20, when he was thanked—through gritted teeth, one imagines. At least the meeting wasn't as dull as usual. One of those present wrote to the *Times* to ask if it was appropriate for the president to use the phrase "God in Heaven knows," and John Herschel, a close friend to whom Babbage had shown a draft, said privately that he'd like to slap Babbage's face. Babbage stopped attending the Society's meetings.

There was more to come. In November, the astronomer Sir James South published a savage little pamphlet titled *Charges Against the President and Council of the Royal Society*. Some of South's thirty-six charges echoed Babbage's complaints. He accused Gilbert and the council of doctoring the minutes of council meetings and suppressing legitimate attempts to reform membership of the Society. But he also charged them with showing disrespect to the king "by presenting to those persons to whom such prizes were awarded, empty boxes in lieu of Royal Medals," and squandering large sums on tavern bills—or, as he picturesquely put it, converting hundreds of pounds of the Society's funds into "whitebait, rose water, and sauterne."[6]

That same month, an anonymous author, calling himself only "one of the 687" Fellows, issued *Science Without a Head: or, The Royal Society Dissected*. We know

now that the author was the Milanese-born Augustus Bozzi Granville, a statistically minded physician who drew up tables of Fellows showing their backgrounds and occupations and whether they had ever contributed to *Philosophical Transactions*. His researches showed that the 63 noblemen who were currently Fellows had contributed nothing between them. Ten bishops had published 9 papers (but they all came from one man, the astronomer John Brinkley, who was also Bishop of Cloyne). The army and navy between them contributed 66 Fellows and 35 papers, although 25 of those came from 2 army captains. Seventy-four clergymen, the traditional occupation for scientific virtuosi, could only manage 8 contributions toward improving natural knowledge, "or 0.108 of a paper each," as Granville helpfully pointed out.[7]

And so Granville's lists went on. Physicians and surgeons had always formed a large proportion of the membership: now there were 100 of them, and they contributed 203 papers, although 109 of them came from one man, the surgeon Sir Everard Home.[*] Sixty-three lawyers made 28 contributions. The remaining 286 Fellows, those who followed no discernible profession (although that didn't mean they were men of

---

[*] After Home's death in 1832, the suspicion arose that most of those papers had been plagiarized from the work of his dead brother-in-law, the surgeon and anatomist John Hunter.

leisure—some were instrument-makers and tradesmen, while others were teachers), made 187 contributions, although 238 of them did nothing at all. The conclusion Granville drew from this analysis was that most of the Fellows were not pulling their weight:

> Few, very few indeed, of the several hundred Fellows classed in the manner I have exhibited them to the public, had not, when elected, or have even at this moment, any pretension to be considered as scientific men—few who could be expected to become useful and valuable members—few who cared for the admission, except as it conferred on them an appellation which it was at one time the custom to look up on as honourable.[8]

The solution, he argued, was to restrict the number of Fellows to, say, 600; and to divide them into classes—mathematics, astronomy, chemistry, and so on—with a fixed quota for each class. The largest would be a "free class" containing 130 men who, while not scientists themselves, were "nevertheless very friendly to science, and anxious to promote or patronize it in some way or other."[9] The moneyed patrons, in other words.

The Royal Society's critics found it hard to agree among themselves. Granville attacked Babbage and South for their "querulous" approach to reform. South attacked Granville for suggesting that the president of the Society needn't be a scientist, and wrote to the *Times* to say that *Science Without a Head* had been written by "A Head Without Science."

All these battles for reform were taking place against the backdrop of an embarrassing battle for the presidency. Davies Gilbert decided to stand down in 1830, and in that undemocratic way that the Society's presidents had, he chose his own successor—Prince Augustus Frederick, Duke of Sussex and brother to King William IV.

There was an uproar, and the campaigns for and against the duke were fought out in public in the run-up to the November election. The surgeon and antiquary Thomas Pettigrew, who was closely involved in the negotiations over the duke's nomination, argued his cause in the columns of the *Times*, claiming that the prince's "patriotic feeling and ardent wishes for the glory and prosperity of his country . . . eminently fit him to fill the chair of the first scientific institution in this country."[10] An anonymous Fellow responded angrily, railing at the lack of consultation over the duke's nomination. "In the present degraded state of science in this country," he roared, "the aspirant to the chair of Newton rests his claim on the recommendation of a single individual [i.e., Gilbert] tired of his seat."[11] The gossip columns predicted that the duke would withdraw, and generally supported the reformers. "There is no royal road to learning," said one. "Why should there be any royal road to the highest honours which can be conferred on learning?"[12] Sir James South's attack on the management of the Society erupted into the middle of this public spat, and the press reported gleefully on an argument at the weekly meeting on November 25. It had ended with South walking out in a rage.

Undeterred by the duke's status, the reformers decided they wanted a scientist to lead them, and at the last minute they persuaded the distinguished astronomer Sir John Herschel to stand against the duke—a move provoking smears, bullying letters, and accusations of ingratitude, since Herschel and his father, William, had benefited so much from royal patronage.*
Herschel was undeterred, and the election on November 30 was close. The Duke of Sussex won by 119 votes to 111, reflecting the divisions within the Royal Society between the old guard and those who wanted science rather than status. This result left a legacy of bad feeling. The *Times* declared that "the first scientific establishment in the empire has obtained a prince, and missed a philosopher," and called on the duke to resign.[13]

---

*William Herschel (1738–1822), astronomer and maker of telescopes, discovered the planet Uranus in 1781, which led to his being appointed court astronomer to King George III. He was elected as a Fellow of the Society in the same year.

As it turned out, the Duke of Sussex proved to be quite a good president during the eight years he remained in office, and an active one as well, at least during the first half of his tenure, before poor health and failing eyesight began to keep him from presiding over the meetings of the council. But the desire to shift the emphasis back to science remained, and in 1846 the reformers finally succeeded in engineering a revision of the statutes. This revision was voted into law on February 10, 1847.

At first sight, the 1847 changes seem slight. Under the old statutes, candidates for Fellowship required a certificate setting out their qualifications and giving the names of three proposers. This certificate was displayed in the meeting room at Somerset House, and if, after ten weeks, no Fellows objected, the candidate was elected. There was no limit to the number of Fellows who could be elected in a single year. It was common to see between 30 and 40 new Fellows elected; of these, scientists were in the minority, averaging less than one-third of the total membership.

The 1847 statutes decreed that a candidate's certificate must be signed by *six* Fellows, at least three of whom should know him personally. Elections were to take place once a year, on June 3, and lists of the candidates' names, along with the names of their proposers and seconders, were printed in the first week of May. Only 15 candidates could be elected in any one year:

the council selected them by ballot from the list and sent a second list, containing their recommendations, to the Fellows, along with a letter naming the day and hour of the election. The effect of all of this was to hand some power back to the ordinary Fellows—although the council still controlled much of the process—and, more importantly, to introduce an element of selectivity. If only 15 candidates could be admitted each year, it would be hard to justify admitting dilettanti while excluding scientists who were distinguished in their respective fields. The number of Fellows dropped dramatically, from more than 750 in 1847 to just over 450 at the end of the century. At the same time, the proportion of scientific Fellows to nonscientific Fellows rose quickly, so that by 1860 they were in a majority for the first time in the history of the Royal Society. Forty years later the number of nonscientists was down to a mere 20. The cap on the number of new Fellows has gradually been raised: at the time of writing it stands at 52, plus up to 10 new foreign members. Candidates must have made "a substantial contribution to the improvement of natural knowledge, including mathematics, engineering science and medical science," and competition is fierce. In 2017 there were around 660 candidates for Fellowship and 90 for foreign membership.

The 1847 revision of statutes marked a fundamental shift in the nature of the Society, which went from being a scientific club to being a learned scientific society. From then on, presidents were distinguished men of science, and if they also happened to be wealthy and aristocratic, as they sometimes were, well, that was a bonus. Since 1915 they have usually been Nobel laureates. The great botanist Sir Joseph Dalton Hooker, who served as president of the Society from 1873 to 1878, established the precedent that presidents should serve for no longer than five years, so that no one could acquire the kind of authoritarian grip that Sir Joseph Banks held for forty-two years.

In 1852, the Society's Somerset House quarters were full to overflowing, principally because of the quantity of books therein. Along with the other chartered scientific societies—the Linnean, Royal Geological, Royal Astronomical, and Royal Chemical Societies—it petitioned the government for more spacious premises. After lengthy negotiations the Society was offered the main building at Burlington House, Piccadilly, on the understanding that the Linnean Society and the Royal Chemical Society would also be at that location. All three moved into their new accommodations in 1857, but only ten years later, the government announced it was giving Burlington House to the Royal Academy of Arts, which was then rather awkwardly sharing space in

Trafalgar Square with the National Gallery. The learned societies weren't going to be evicted from Burlington House, however: instead, they would be accommodated in two entirely new wings. These were completed in 1873, and the Royal Society moved into its new purpose-built premises in the east wing that year.

With a new home, a new, more serious purpose, and a membership that now consisted mainly of scientists—and very distinguished scientists at that—the Royal Society entered a new phase in its history. The geologist and historian of the Society Sir Henry Lyons was clear about the significance of the Victorian reforms: "By the end of the nineteenth century the Society, after overcoming much opposition and indifference, had realised the aims of its founders, and at last had become an institution for promoting natural science."[14]

9

FOREIGN PARTS

"Discovering lands towards the South Pole"

FROM ITS FOUNDATION THE ROYAL SOCIETY HAD AN international dimension. It continually sought information on foreign parts through its correspondence networks, and Fellows often asked sea captains and other travelers to keep an eye out for curiosities of nature and exotic flora. These requests were very fruitful. On February 10, 1670, for example, members were entertained with a letter from John Winthrop, the governor of Connecticut and the Society's first colonial member, who had sent over a parcel containing forty-one items of interest, ranging from examples of Native American wampum and "a curious sort of moss growing on the trees beyond Virginia" to a pair of flying squirrels and some ears of Indian corn.[1]

Sometimes Henry Oldenburg supplied sympathetic travelers with a detailed questionnaire to take with them on their journeys. After the Hudson's Bay Company was founded in 1670, the Royal Society secretary gave Zachariah Gillam—a New England sea captain who commanded one of the company's first expeditions to East Hudson's Bay—a long list of questions to answer on matters such as tides, magnetic variation, and the habits of beavers. Gillam was able to provide comprehensive answers. Of the indigenous people they encountered, he said they were nomadic, lived "to an hundred and twenty years," and drank venison broth. They were also given to taking what we would now call saunas:

> Concerning their physic, they use chiefly sweating, not by taking any thing inwardly, but by making a kind of stove, heating many stones red hot, in their tents, and then pouring water upon them, whereby they are made to sweat excessively, in which condition, when they have sat a while, they run out into the snow, whereby they say their pores are presently closed again, as they were opened by the heat.[2]

The next step for the Society was the sponsorship of expeditions. Between 1698 and 1701, the astronomer

Edmond Halley, who would later predict the reappearance of the comet that still bears his name, made three voyages as captain of the HMS *Paramore*, with the aim of improving "the knowledge of the longitude and variations of the compass."[3] These were Royal Navy expeditions—and Halley was commissioned as a full captain in the king's navy—but they had the official support of the Society and rank as perhaps "the earliest sea journeys undertaken for a purely scientific object."[4]

Fellows frequently participated in scientific expeditions to far-flung places or were present on military or commercial voyages that enabled them to make scientific observations: Hans Sloane, who went to the West Indies in 1687 as physician to the new governor of Jamaica, later published a catalog of Jamaican plants. When he was only twenty-three, Joseph Banks sailed aboard the frigate HMS *Niger*, which was patrolling the fisheries off Newfoundland and Labrador from 1766 to 1767.

The first major scientific expedition to be sponsored by the Royal Society (albeit with considerable assistance from the Royal Navy) set sail at the end of the same decade. In 1761, the Society had sent Nevil Maskelyne to the island of Saint Helena in the South Atlantic to observe the transit of Venus across the sun—a phenomenon that had interested scientists for more than a century, because it offered them a better means of estimating the distance between the earth and the sun. Low clouds meant that Maskelyne's journey was not a great success, but another transit—the last for 105 years—was expected in 1769. Throughout the 1760s the Society lobbied for expeditions to observe the 1769 transit in both hemispheres. A voyage to the South Seas seemed particularly advantageous, not only because of the transit but also because it presented the possibility of making "a settlement in the great Pacific Ocean, or . . . discovering lands towards the South Pole."[5] A committee was set up in 1766 "to send astronomers to several parts of the world to observe the transit," and two years later, the Society urged George III to send ships both north and south, for the sake of Britain's reputation and standing in the world. Observers, the Society recommended, should be sent to Hudson's Bay in Canada, to the North Cape in Norway, and to "any place not exceeding thirty degrees of southern latitude, and between the 140th and 180th degrees of longitude west from your majesty's royal observatory in Greenwich Park."[6] The cost was estimated at about £4,000, excluding the price of the ships themselves. The southern expedition would need to leave by the summer of 1768 if it was to be in the best position to make observations when the transit occurred on June 3, 1769. The document drawn up for the king emphasized that the Society was "in no condition to defray this expense," as its annual income was "scarcely sufficient to carry on the [Society's] necessary business."[7]

The king agreed to a personal grant of £4,000, and the Admiralty bought a 368-ton Whitby collier for the southern voyage, the *Earl of Pembroke*. Joseph Banks saw an opportunity to explore the natural history of the South Pacific and offered to finance a subsidiary group to accompany the expedition. In May 1768, by which time the *Earl of Pembroke* had been renamed the *Endeavour*, James Cook was appointed as ship's commander.

Captain Cook and the Society's appointee as astronomer, Charles Green, duly made their observations at Tahiti. But the voyage of the *Endeavour* had additional significance. After the transit, Cook followed what were called his "additional instructions," sailing south and then west in search of the great continent that was supposed to exist in the Southern Ocean. He surveyed the north and south islands of New Zealand and the east coast of Australia, which he took possession of in the name of George III and named New South Wales. Botany Bay was named for the great quantity of plants that Banks and his team discovered there. The *Endeavour* returned to England in June 1771.

In spite of Cook's "additional instructions" and their imperialist subtext, the voyage of the *Endeavour* remains the Royal Society's greatest scientific expedition. It placed the Society at the heart of one of the eighteenth century's great voyages of discovery, while Banks's own observations on the botany, zoology, and ethnology of the South Pacific were of lasting significance.

In the nineteenth century, the Royal Society continued to support—and sometimes to take credit for—the work of individual members in the field of scientific exploration. Edward Sabine (1788–1883), the pioneering crusader in the search for the causes of geomagnetism, served as secretary, foreign secretary, treasurer, vice-president, and eventually president of the Society. He was instrumental in promoting James Clark Ross's great voyage of exploration to the Antarctic in 1839–1843 in the strengthened ice-ships *Erebus* and *Terror*, and helped persuade the Royal Society to support the expedition. He was also successful in enlisting the Society's backing for the establishment of a chain of geomagnetic observatories across the empire, manned by military and naval personnel but essentially under Sabine's control.

Between 1819 and 1891 the Society supported nine expeditions, three apiece to the Arctic, the Antarctic, and the continent of Africa. All yielded important scientific results. Some were enormously significant, such as the voyage of the steamship *Challenger*, which in 1872 set off on a mission to research marine zoology and oceanography that took it to Cape Town, into the Antarctic Circle, and to New Zealand and the Pacific before it passed through the Straits of Magellan and headed for home. All the while its crew took water samples and retrieved marine life, using metal nets to reach specimens from the sea floor at depths of up to 3,000 fathoms. The results of the *Challenger*'s researches were so prolific—and so critical to the development of the modern science of oceanography—that the Society set up a special committee to oversee their publication, a process that continued over the next twenty-four years.

But Joseph Banks had been right to worry about competition from the newer societies. An expedition to observe the 1874 transit of Venus was carried out under the auspices of the Royal Astronomical Society. An initiative in the 1890s to promote a National Antarctic Expedition came from the Royal Geographical Society rather than the Royal Society, and although the two institutions set up a joint organizing committee, their representatives fell out with each other. The Geographical Society went ahead in 1901 with its own candidate, a young Robert Falcon Scott, in sole charge of what became known as the *Discovery* expedition. The Royal Society did give Scott and his team a splendid dinner before their departure for the Antarctic, and in return, when he arrived in McMurdo Sound aboard the *Discovery*, Scott named the high western mountains he saw there the Royal Society Range. Ernest Shackleton, a member of Scott's 1901 expedition, didn't even bother to seek funding from the Society for his own British Antarctic Expedition of 1907–1909 (he asked if he might borrow a magnetometer, but the Society turned down his request because the instrument was already committed elsewhere).

Expeditions with a primary focus on exploration fell outside the Society's remit, although it often played a crucial role in giving scientific advice and moral support to those who did carry them out. Such expeditions often included personnel who were—or who would become—Fellows: Shackleton's 1907 expedition, for example, included the professor and Royal Society Fellow T. W. Edgeworth David, who reached the South Magnetic Pole and climbed to the top of Mount Erebus. And no fewer than five members of a team that researched overfishing around the Falkland Islands in the 1920s were later elected Fellows.

The Royal Society mounted its own expeditions only rarely. In 1936, a team sponsored by the Society spent four months on the island of Montserrat, making seismological observations to serve as a basis of comparison in the event of any serious changes in the volcano dominating the island. And in the 1950s, the Society established a geophysical observatory in Antarctica as part of the United Kingdom's contribution to the International Geophysical Year (which actually lasted for eighteen months, from July 1, 1957, to December 31, 1958), with funding from the Treasury. The bay where the advance party landed was named Halley Bay in honor of Edmond Halley, whose tercentenary had been celebrated in 1956. A second postwar expedition followed in 1958. This time the Society mounted a small expedition to southern Chile, with the aim of investigating relationships between the biology of New Zealand and the southern tip of South America.

As a result of the experience gained in mounting these two expeditions, in 1959 the council approved the establishment of a scientific expeditions advisory service and an expeditions department. For the next thirty or forty years the Society ran a series of overseas projects, sometimes alone and sometimes in partnership with other institutions. In 1965 it sent an ambitious biological expedition to the Solomon Islands, and in the 1980s it embarked on a long-running research program in the rainforests of northern Borneo. Geologists studied volcanic activity on Tristan da Cunha, and biologists established a research station on the coral atoll of Aldabra to study the breeding ground of the giant tortoise inhabiting the region of the Indian Ocean.

This fine flowering of an impulse that began with Halley back in the seventeenth century slowed down at the end of the twentieth as the idea of the expedition became ever more complex, multidisciplinary, and ideologically fraught. In the 1990s, the Society's expeditions department was quietly wound up, although research expeditions by individual Fellows continue today.

10

A BRAVE NEW WORLD

*"They wish to see scientists doing more
to help the world in its difficulties"*

THE TWENTIETH CENTURY ARRIVED ON THE ROYAL
Society's doorstep with a crash. On May 10, 1900, the
council recorded that the naturalist Marian Farquharson
had written to suggest that "duly qualified women
should have the advantages of full Fellowship."[1] Farquharson had campaigned for women to have full access
to a range of learned societies. She had already been
admitted as the first female Fellow of the Royal Mi-
croscopical Society in 1885, although it was something
of a pyrrhic victory, since she wasn't allowed to attend

any of its meetings. The Royal Society didn't even go that far: it fobbed her off by maintaining that the admission of women depended on the interpretation placed on the royal charters "under which the Society has been governed for more than three hundred years," and there the matter rested.[2] But two years later, in January 1902, they encountered a more direct challenge when the Fellow John Perry proposed Hertha Ayrton, a distinguished electrical engineer, for membership. Perry, who was also an engineer, supported his proposal with a list of distinguished cosignatories. It included astronomers, physicists, chemists, and engineers, and they all knew Ayrton personally.

Although no woman had ever been elected a Fellow, women had figured in the Royal Society's history from its earliest days. In May 1667, Margaret Cavendish, Duchess of Newcastle, a writer with wide-ranging intellectual interests, was entertained with her ladies at Arundel House with various experiments, including the weighing of air. Samuel Pepys, who was at the meeting, was not impressed, complaining of her "antic" dress and the fact that she said nothing "that was worth hearing, but . . . was full of admiration, all admiration." In 1832 the Society admitted the science writer Mary Somerville—but only in effigy. Her bust, carved by Francis Leggatt Chantrey, was placed in the hall at Somerset House.

Women figured more substantially in the *conversazioni* that were an important part of the Royal Society's social life by the later nineteenth century. These evolved from the informal receptions hosted by Sir Joseph Banks at his home and at his own expense. The Society took over responsibility for them in 1871, forming a soirée committee with its own small budget. The committee planned the receptions thereafter and provided the wine, ices, and musical entertainment. They took place in May and were initially male-only events, although the work of at least two women was exhibited: photographs of early Christian architecture in Ireland, taken by the archaeologist Margaret Stokes, and paintings of rare plant life made by the naturalist Marianne North.

Then in 1876 the soirée committee introduced a second *conversazione* in June, to which the president invited both men and women. "The June reception," gushed the *Times* in 1899, "is graced by the presence of the other sex, whose variegated adornments impart an unwonted gaiety to the severe environment of the headquarters of British science."[3] The "ladies' *conversazioni*," as they were called, usually recycled the exhibits and experiments that had been shown to the all-male audience the previous month; but in June 1899, along with photographs of Patagonia, a horizontal-pendulum seismograph, and a magic lantern show of an erupting Vesuvius, Hertha Ayrton personally demonstrated her experiments with the electric arc. "For more reasons than one Mrs Ayrton's experiments . . . attracted considerable attention," noted the *Times*. The following year, John Perry read Ayrton's paper on "the mechanism of the electric arc" to a full meeting of the Society.

Although Ayrton's work was already known to the Society, Perry's proposal put the council into a panic. Legal advice was sought, and King's Counsel came back with the opinion that married women were ineligible for election, since in common law a husband and wife were one person, and that person was the husband. The wife had no separate legal existence. Because Hertha Ayrton was married to the physicist, electrical engineer, and Royal Society Fellow William Ayrton, she was automatically barred.

The rebuff didn't prevent Hertha Ayrton from becoming involved in the Royal Society. She read a paper on wave motion at the Society in June 1904, becoming the first woman to present at a regular meeting, and in 1906 she received the Society's Hughes Medal. This medal was awarded annually for original discoveries in the physical sciences, and the Society chose Ayrton in recognition of her work on the electric arc and on sand ripples. "Can we now refuse the Fellowship to a Medallist?" asked the worried past president and current council member William Huggins, who disapproved strongly of her being given the award.[*]

The Royal Society at the time was divided into factions for and against the admission of women, with a large group in the middle that didn't quite know what to make of the idea. They weren't alone in this. A number of learned societies excluded women, either by statute or simply by refusing to let them in. The Royal Academy, for example, had had two female painters, Angelica Kauffman and Mary Moser, among its founding members in 1768, so there was technically no reason to bar women, and yet no more women were admitted until Laura Knight came along in 1936. The times were changing, though. The Linnaean Society admitted its first women Fellows in 1905, largely as a result of campaigning by Marian Farquharson. The Royal Geographical Society decided to admit women in 1913, following a debate that had lasted for twenty

---

[*] At the time of writing, Ayrton is one of only two women to be awarded the Hughes Medal. The other is Michele Dougherty, who won the Hughes in 2008 for her "innovative use of magnetic field data that led to discovery of an atmosphere around one of Saturn's moons and the way it revolutionised our view of the role of planetary moons in the Solar System." "Michele Dougherty: Biography," Royal Society, https://royalsociety.org/people/michele-dougherty-11354.

years; and two years later, the Royal Astronomical Society admitted its first women Fellows, having previously argued that because its royal charter of 1831 referred to Fellows as "he," it was against the rules.

Ironically, the chemist Henry Armstrong, a close family friend of the Ayrtons, was one of the loudest opponents of women members. Will Ayrton, he wrote, "should have had a humdrum wife who would have put him into carpet-slippers when he came home, fed him well and led him not to worry . . . ; then he would have lived a longer and a happier life and done far more effective work."[4] With attitudes like this, it isn't surprising that progress toward equality was slow—as it was across the entire scientific community. Hertha Ayrton never was elected a Fellow, but she did become active in the suffragette cause, demonstrating in Downing Street with Emmeline Pankhurst in 1910. She provided her home as a place for Pankhurst and other hunger strikers to regain their strength, and when her daughter Barbara was sent to prison for smashing windows in 1912, she wrote, "Barbie is in Holloway . . . I am *very* proud of her."[5]

After World War I and the triumph of suffragism came the 1919 Sex Disqualification (Removal) Act, which stipulated that "a person shall not be disqualified by sex or marriage from the exercise of any public function, or from being appointed to or holding any civil or judicial office or post, or from entering or assuming or carrying on any civil profession or vocation, or for admission to any incorporated society (whether incorporated by Royal Charter or otherwise)."[6] It made no difference: John Perry and all but one of his cosignatories were dead by 1919. Hertha Ayrton died four years later.

No woman was proposed for membership again until World War II, when the biochemist Marjory Stephenson and the crystallographer Kathleen Lonsdale were both put forward. After a vote on amending the statutes showed an overwhelming majority in favor, the Society elected Stephenson and Lonsdale as members on March 22, 1945, and they became the first female Fellows of the Royal Society.

Today, 5 percent of the living Fellows and foreign members are women.

❧

While the struggle for women's rights left most Fellows unmoved, many of them reflected on their own roles and on the function of the Royal Society in a brave new world of poison gas and penicillin, where science promised to be both the servant and the master of humanity.

One effect of this process of reflection was a reinforcement of the Society's function as broker, adviser, and sponsor. It played a crucial part in the establishment in 1900 of a National Physical Laboratory "for standardising and verifying instruments, for testing materials, and for the determination of physical constants."[7] The NPL, one of the oldest standardizing laboratories in the world, was under the control of the Society's president and council for the first eighteen years of its existence. The Society also represented Britain in the new International Research Council set up in 1919 to promote international cooperation in the sciences.

Between 1919 and 1923, the Royal Society attracted bequests amounting to more than £350,000, allowing it to create four professorships and an important research laboratory at Cambridge. The Nobel Prize winner Ernest Rutherford, the father of nuclear physics, who served as president from 1925 to 1930, took care to bring the Society close to the heart of government: during his tenure, two prime ministers, Stanley Baldwin and Ramsay MacDonald, were elected Fellows under a Society statute allowing the council to recommend the election of men who had rendered conspicuous service to the cause of science "or are such that their election would be of signal benefit to the Society."[8] Rutherford also did his best to inject new life into the regular meetings. Henry Armstrong approved of the change wholeheartedly, telling Rutherford in his customary tongue-in-cheek way that "your attitude in the chair is delightful: to have a president asking questions and promoting discussion is an astounding departure."[9]

However, under Rutherford's successor, another Nobel Prize winner, the biochemist Sir Frederick Gowland Hopkins, there were murmurs of dissent among the Fellows about the way in which a cabal—essentially the president, officers, and council—was keeping the Society too close to the establishment, instead of taking a lead in promoting social responsibility in the sciences. The dissent was led by Frederick Soddy, yet another winner of the Nobel Prize. There have been more than 280 Nobel laureates among the Fellows and foreign members since the Nobel Foundation began awarding prizes in 1901, including such titans of modern intellectual inquiry as Albert Einstein, Max Planck, and Francis Crick and James Watson, who—along with Rosalind Franklin and others—identified the structure of DNA. Himself a distinguished chemist, Soddy dabbled in some of the many individualist economic fringe movements that sprang up in response to the Depression in the early 1930s, rejecting the twin peaks of capitalism and collectivism and trying to forge a path between them. Soddy wanted the Royal Society's ordinary Fellows to play a more active role in determining the strategic direction of the Society: as it was, he claimed, the cabal suppressed those aspects "not favoured by or dangerous to authority, whether scientific or public."[10] The problem was that a habit had grown up over the years by which the old council nominated the president, the officers, and the new council, and these nominations were confirmed without contest by the Fellows. Soddy argued for a postal ballot, which would engage more Fellows in the process, and ninety-one of the Fellows supported him. When the council heard of his concerns, it responded that this was the way things had always been done: the dissenters, it declared, "failed to realise the revolutionary changes in the constitution of the Society which some of its wording implied."[11]

In fact, they realized them perfectly well. At the anniversary meeting of 1935, Hopkins stood down at the end of his five-year term, and Soddy and his comrades fielded a rival candidate for president, the immunologist Sir Almroth Wright. Soddy stood for treasurer, and the eight vacant seats on the council were all contested by reformers. An extraordinary half of all Fellows, including Rutherford and MacDonald, turned up to vote on November 30, 1935—and the conservatives swept the board. But as the *Manchester Guardian* reported afterward, the contest suggested that a significant minority of Fellows nevertheless sympathized with the reformers. Although they weren't prepared to oust the old guard, they wanted the Society to provide "a more positive lead concerning those contemporary social problems that have definite scientific aspects," the *Manchester Guardian* said, adding, "They wish to see scientists doing more to help the world in its difficulties."[12]

Evidence for this came in Sir Frederick Hopkins's outgoing presidential address. He spoke on the social responsibility of scientists and the difficulties of controlling the social effects of scientific discovery. And in a placatory nod to the dissidents in the room, he declared that although in the past governments had frequently sought advice through the Royal Society, now the Society "should no longer consider its part restricted to advice, but, on occasion when it had something definite and important to say, should take the initiative."[13]

⟨≈⟩

The question of the moral role of the scientist remains as relevant today as it was in 1935. More so, perhaps. In a recent anniversary address, the current president of the Society, the biologist Venki Ramakrishnan, argued that the Fellows "need to ensure that we remain strong advocates for the right sorts of decisions for science and more generally for the country," a statement that would have warmed the hearts of Soddy and his friends and astonished his predecessors in the eighteenth and nineteenth centuries.[14] The battle for experimental philosophy has long since been won: now the Royal Society announces itself to be "the independent scientific academy of the UK and the Commonwealth, dedicated to promoting excellence in science."[15] Its mission, besides promoting excellence, is to "encourage the development and use of science for the benefit of humanity."[16]

In the twenty-first century, the club founded by those twelve experimental philosophers in Lawrence Rooke's lodgings back in 1660 has evolved into a more public-facing, reflective organization engaging with a whole series of issues, from climate change and embryo research to genetically modified foods and data governance. Having outgrown the east wing of Burlington House, in 1967 it moved to its current premises in Carlton House Terrace, where it maintains forty-one standing committees and six working groups. It investigates cybersecurity, mathematics education, and diversity along with more traditional scientific topics and advises the government in these areas. It awards research grants to early career scientists, and it runs an entrepreneur-in-residence scheme to introduce university staff and students to state-of-the-art industrial research. At the time of writing in 2018, the Royal Society funds 1,500 researchers and administers 29 medals and awards. They range from the prestigious Copley Medal for outstanding achievements in scientific research, first awarded in 1731, and thought to be the world's oldest scientific prize, to the Athena Prize, established in 2016 and awarded to the individuals and teams that have contributed the most to the advancement of diversity in science, technology, engineering, and mathematics. There are now around 1,600 Fellows and foreign members in the Royal Society, who are supported by a staff of 160. The *Philosophical Transactions*, *Notes and Records*, and *Proceedings* are all going strong.

Does the Royal Society of London still have a role to play in a world where governments, academic institutions, and private businesses determine so much of the course of scientific research and control its benefits? Of course it does, and precisely because so many vested interests are bent on manipulating scientific advances in their own interest. An independent voice in the scientific community is more important now than it has ever been. There was so much to learn in 1660. There is more to learn today.

Nullius in verba: Take no one's word for it.

APPENDICES

APPENDIX 1:
THE FOUNDERS

TWELVE MEN WERE PRESENT AT THE INAUGURAL
meeting of the Royal Society at Gresham College on
November 28, 1660. They came from different backgrounds and had different political outlooks. Some
were royalists, some republicans. Some were career
scientists and academics, while others were dabblers.
What united them was a commitment to a Baconian
conception of how to arrive at knowledge—by experiment, by investigating things for oneself rather than
accepting ancient authority.

WILLIAM BALL (C. 1631–1690), ASTRONOMER

Ball came from a family of minor gentry with estates at Mamhead in Devon. He trained as a lawyer, entering the Middle Temple in 1646, although there is no evidence that he ever practiced law. As far as we know, his interest in astronomy dates from the mid-1650s, when, like Sir Paul Neile, Christopher Wren, and others, he was making observations of Saturn in an attempt to explain its changing appearance. By the end of the 1650s he was a regular attendee at the lectures held in Gresham College, and at the first meeting of the new college "for the promoting of physico-mathematical experimental learning" he was appointed treasurer. At the first annual election in 1663, that post went to Abraham Hill, although Ball was elected to the council.

In 1665 Ball retreated to his father's estate in the West Country to avoid the plague. He returned to London briefly after the Great Fire and married Mary Posthuma Hussey at St. Paul's Covent Garden in July 1668. They settled in Devon, where they had six children. Ball spent the last third of his life managing the family estates, a career that left him little time for science.

ROBERT BOYLE (1627–1691), NATURAL PHILOSOPHER

The youngest son of Richard Boyle, 1st Earl of Cork, Robert Boyle went to Eton before being sent on a Grand Tour of Italy in 1641. His father left him an estate at Stalbridge in Dorset, and in 1645 he moved there, although he also spent time in London, where he became involved with one or more of the groups of experimental philosophers active in the capital in the later 1640s. In 1649 he established a laboratory at Stalbridge, where he conducted a series of chemical and alchemical experiments, his interest in experimentation growing all the time. In 1655 he moved to Oxford to join John Wilkins's Wadham College group. There, he employed the young Robert Hooke to assist him with his experiments. Throughout the early 1660s Boyle published a stream of important scientific works; when his eyesight began to fail, he continued with the aid of an amanuensis who took down his dictation. His *New Experiments Physico-Mechanical, Touching the Spring of Air and Its Effects* (1660) described his groundbreaking experiments with a vacuum pump; his response to criticism of this work, *A Defence of the Doctrine, Touching the Spring and Weight of Air* (1662) set out his theory, known to posterity as Boyle's Law, that the volume of a constant mass of ideal gas at a constant temperature is inversely proportional to its pressure.

Boyle was beyond doubt the most distinguished scientist among the first twelve members of the Royal Society. An ardent Protestant (he turned down the presidency of the Royal Society because he believed the taking of oaths to be unbiblical), he was also actively involved in plans to convert Native Americans to Christianity. After a stroke in 1670, his direct involvement with the Royal Society slowed, although the laboratory he established at the house in Pall Mall that he shared with his sister Lady Ranelagh from 1668 onward was a place of pilgrimage for visiting scientists. He died on New Year's Eve, December 31, 1691, one week after the death of his sister, and was buried in the chancel of St. Martin-in-the-Fields.

WILLIAM BROUNCKER, 2ND VISCOUNT BROUNCKER (1620–1684), MATHEMATICIAN

The first president of the Royal Society, Brouncker was born in Dublin County, the son of a minor courtier. After qualifying as a physician at Oxford in 1647 he turned to mathematics, translating Descartes and exchanging ideas with John Wallis, Savilian Professor of Geometry at Oxford. He was the first to express the ratio of the area of a circle to the circumscribed square as an infinite continued fraction. His studies earned him a reputation as a mathematician, and his relationship with Wallis led him to Gresham College in the period immediately before the Restoration. Brouncker was named as one of those who used to meet at Gresham in 1659, either after Wren's astronomy lectures on Wednesdays, or after Rooke's geometry lectures on Thursdays. Two weeks after the inaugural meeting of the Society on November 28, 1660, Brouncker was part of a committee, along with Moray, Goddard, Neile, and Wren, who were charged with finding a convenient place for the weekly meetings of the new society. He was a regular attendee and served on several other committees in the first year, including one with the rather curious dual purpose of "erecting a library, and examining the generation of insects."

Brouncker was named president of the Royal Society in the first charter of 1662, and this position was confirmed in the second charter of 1663. He was reelected without a contest every year until 1677, by which time he was losing interest in the Society and missing meetings. When a faction in the Society proposed Sir Joseph Williamson as a candidate and it became clear there would be a contested election, "Lord Brouncker in great passion raved and went out."[1] He didn't turn up on St. Andrew's Day, and Williamson was duly elected as the Royal Society's second president. Brouncker died on April 5, 1684, at his house in St. James's Street Westminster.

ALEXANDER BRUCE, 2ND EARL OF KINCARDINE (1629–1681), LANDOWNER

The son of a Scottish nobleman with stone and marble quarries and coal mines on the family estate at Culross, Bruce left Scotland and went into exile in Bremen in 1657, presumably because of his adherence to the Royalist cause. From there he moved to Hamburg, collaborating with Christiaan Huygens on an attempt to invent an accurate pendulum clock that could determine longitude at sea. He was also in regular correspondence with fellow Scot Sir Robert Moray, himself in exile in Maastricht; letters between the two men show they shared interests in chemistry, physics, and mathematics, among many other subjects. Bruce returned to Culross at the Restoration, although, like many other exiled Royalists, he was also drawn to Charles II's court at Whitehall in the hope of preferment. This, and his links with Moray, may explain his presence at the inaugural meeting at Gresham.

Bruce inherited the Kincardine earldom and the family estates in 1662, and thereafter his involvement in Scottish politics meant that he played only a peripheral role in Royal Society affairs. After Bruce's death on July 9, 1680, the Scottish philosopher and historian Gilbert Burnet recalled that his "thoughts were slow, and his words came much slower; but a deep judgement appeared in everything he said or did."[2]

JONATHAN GODDARD (1617–1675), PHYSICIAN

In 1646, after studying medicine at Oxford and Cambridge, Goddard, the son of a prosperous shipbuilder, was elected a Fellow of the College of Physicians. While developing a successful general practice, he indulged his scientific interests by joining the loose grouping of experimental philosophers, many of them also physicians, who gathered in London in the mid-1640s. According to John Wallis, one of the places they gathered was Goddard's Wood Street lodgings, because he kept an operator for grinding lenses there. After a spell in Ireland and Scotland as physician-in-chief to Cromwell's army, he was rewarded with the wardenship of Merton College Oxford in 1651, and four years later he was appointed professor of physic at Gresham. Thus, Goddard was ideally placed to engage with both the Wadham group under Wilkins and the later Gresham group.

He appears regularly in the early records of the Royal Society as serving on various committees: one for "the experiments about shipping," for example, with Petty and Wren, and another "for considering of proper questions to be inquired of in the remotest parts of the world."[3] Not surprisingly, considering his background, he was particularly interested in anatomy and physiology. He researched the nature of respiration and worked with Boyle on the compression of air in water. On one occasion he gave a paper on observations he had made "in the dissection of a chameleon." Goddard died on the corner of Wood Street on March 24, 1675, after suffering a stroke while on his way home from a meeting of natural philosophers.

ABRAHAM HILL (1633–1722), MERCHANT

Hill was the son of a prosperous London alderman. His first wife, Anne, was daughter of the prominent English lawyer and politician Sir Bulstrode Whitelocke. When Hill's parents died in 1660, he bought a country house in Kent and took rooms in Gresham College, where he came to know the other founders of the Royal Society. Anne died the following year, and Hill remarried shortly thereafter. A dabbler rather than a serious scientist, Hill was nevertheless a capable administrator, and he played an active role in the Society for more than half a century. He was a member of the council from 1663 to 1666, and again for the forty-nine years from 1672 to 1721, the year before his death. He served as secretary from 1673 to 1675 and did two stints as treasurer, from 1663 to 1666 and from 1677 to 1699. He also helped to plan experiment schedules for Society meetings and to prepare the long question-naires the Society was wont to give to sea captains and other travelers.

Hill died at his Kent estate on February 5, 1722.

SIR ROBERT MORAY (C. 1608–1673), SOLDIER AND COURTIER

Moray was the son of Sir Mungo Moray of Craigie in Perthshire. Like many Scots of his generation, he spent most of his early life soldiering in Europe in the Thirty Years' War. He was in England in January 1643, when his loyalty to the Royalist cause earned him a knight-hood from Charles I, but soon afterward he was back in France as lieutenant-colonel of a new French reg-iment, the Scottish Guards. However, he remained active in the king's cause, taking part in unsuccessful negotiations to bring the Scots into the English Civil War on the side of the Royalists. After Charles I's exe-cution, Moray continued to work for the Restoration of Charles II. In 1655 he was forced into exile and stayed with the court at Cologne and Bruges before moving to Maastricht in 1657. There is little evidence that he had more than an amateur's interest in scientific matters, but it was enough for him to make the acquaintance of the Gresham group.

At the Restoration, when he was back in London and being heaped with honors for his devotion to the Royalist cause, he was at Gresham for the inaugural meeting of the Royal Society with Alexander Bruce. He proved a good friend to the Society, engineering the grant of a royal charter and consistently promoting the Society to Charles II. John Aubrey recalled that Moray "was a good chemist and assisted his majesty in his chemical operations."[4]

SIR PAUL NEILE (1613–1686), COURTIER

Neile had a solid Anglican background: his father was bishop of Lichfield and later archbishop of York, and his wife, Elizabeth Clarke, was the daughter of the archdeacon of Durham. After a rather wild youth, in the course of which his father had to intervene to help him avoid prosecution for killing a cart driver, he seems to have had a quiet Civil War, in spite of his strong royalism. It was during the Commonwealth years that he forged strong working relationships with some of the Wadham set: his son William, later to become a distinguished mathematician, went up to Wadham in 1652. Neile was known for his interest in optics, paying for the construction of a series of powerful telescopes, one of which he later gave to his friend Christopher Wren when Wren was appointed to the astronomy chair at

Gresham in 1657. It was demonstrated to Charles II in October 1660, and the king was so impressed that he asked for it to be installed at Whitehall.

Neile, named as a member of the council in the first charter and again in the second, was made a gentleman usher of the privy chamber to Charles II in June 1662, and he used his proximity to the sovereign to further the interests of the Royal Society whenever he could. He was a regular at the weekly meetings of the Society, reading a discourse on cider at one in July 1663, and reporting on another occasion that he had seen an eel with a duck in its mouth in St. James's Park. Little is known of his later life, although he seems to have been living at Codnor Castle in Derbyshire in the early 1680s.

Sir William Petty (1623–1687), Political Economist and Physician

After going to sea as a boy and ending up in France, Petty entered the Jesuit college at Caen in about 1637, where he studied mathematics. He returned to England to continue his education, but on the outbreak of war he went back to Europe, studying at Amsterdam, Leiden, Utrecht, and Paris before entering Oxford in 1646. His aim by then was to qualify as a physician, which he duly did in 1650. After the ejection of Royalist academics from Oxford, Petty, whose political principles moved with the times, was appointed vice-principal of Brasenose College and professor of anatomy. In 1651, with that cheerful absence of specialization that was such a feature of the early experimental philosophers, he became professor of music at Gresham College before joining Cromwell's army in Ireland as a military physician. He then embarked on a massive survey of forfeited lands there, acquiring quite a lot of them for himself.

Although Petty was forced to relinquish some of his Irish estates at the Restoration, he moved back and forth between England and Ireland for the next quarter century and engaged in a series of ventures, from designing a catamaran to pioneering statistical analysis. He had a formidable intellect and a deep commitment to experimental philosophy. John Aubrey called Petty "a person of an admirable inventive head,"[5] and Samuel Pepys said he was "the most rational man that I ever heard speak."[6] When Aubrey suggested that the Royal Society should schedule its annual elections for St. George's Day instead of the feast of St. Andrew, patron saint of Scotland, Petty replied that he would rather have the elections on St. Thomas's Day, because the apostle Thomas, known as Doubting Thomas, "would not believe until he had seen and put his fingers into the holes" in the hands and side of Christ left from the crucifixion.[7]

LAWRENCE ROOKE (1622–1662), ASTRONOMER

Rooke's early life was uneventful. He went to Eton and then in 1639 was admitted to King's College Cambridge, where he studied for the next eight years. After a spell at the family home in Kent, he entered Wadham in 1650, apparently so that he could work with the warden, John Wilkins, and Seth Ward, who was then Savilian Professor of Astronomy at Oxford. In 1652 Rooke got the astronomy chair at Gresham, and in 1657 he swapped it for the chair in geometry, reputedly because the geometry professor's lodgings were quieter and more comfortable. It was in those lodgings that the inaugural meeting of what would become the Royal Society took place. Like most astronomers of the day, Rooke was particularly interested in astronomical solutions to the problem of longitude. He delivered a paper on lunar eclipses to an early weekly meeting of the Society. He also wrote about the eclipses of Jupiter's moons and longitude.

Rooke's career came to an untimely end in June 1662, before the grant of the charter to the Royal Society was made, so he was never technically a Fellow. He contracted a chill on the way home to Gresham after a visit to his patron, the Marquis of Dorchester, and died on the night of June 27 at the age of forty. On his deathbed, he said that he had only one more observation to make of the moons of Jupiter, and begged his colleagues to make it for him. He was remembered as "profoundly skilled in all sorts of learning, not excepting botany and music, and the most abstruse points of divinity; [and] always averse from asserting any thing positively, that was dubious."[8]

JOHN WILKINS (1614–1672), NATURAL PHILOSOPHER

If anyone had a claim to be the driving force behind the formation of the Royal Society, it was Wilkins. An ordained priest from a family of moderate Puritans, he published a stream of speculative scientific works between 1638 and 1648, beginning with *The Discovery of a New World, or, A Discourse Tending to Prove, That ('Tis Probable) There May Be Another Habitable World in the Moon*, and ending with *Mathematical Magick, or, The Wonders That May Be Performed by Mechanical Geometry*. He was at the heart of the scientific circles that met in London in the mid-1640s, and when he went to Oxford in 1648 to take up the post of warden of Wadham College, a number of other experimental philosophers moved there as well, drawn by the informal meetings and demonstrations that took place under his watchful eye. Aubrey called him "the principal reviver of experimental philosophy . . . at Oxford."[9] This group included Seth Ward, who became Savilian Professor of Astronomy, Lawrence Rooke, Jonathan Goddard, William Petty, and Robert Boyle. Wilkins maintained a reputation for tolerance, with "nothing of bigotry, unmannerliness, or censoriousness, which then were in the zenith, amongst some of the heads, and Fellows of colleges in Oxford."[10] He attracted brilliant young students from all backgrounds, including Christopher Wren, William Neile, and Thomas Sprat. His marriage to Oliver Cromwell's widowed sister, Robina, in 1656 did his prospects no harm, and in 1659 he was made master of Trinity College Cambridge.

The Restoration brought Wilkins's university career to an end, and he moved back to London, where he again became the focus for scientific activity. He chaired the inaugural meeting of the Society, and he became a member of the council and one of the Society's two secretaries, along with Henry Oldenburg. He also collaborated with Sprat on the latter's *History of the Royal Society*, helping to establish and confirm the Society's emphasis on the experimental method.

Wilkins lost his house and all his papers in the Great Fire of 1666. In 1668 he was consecrated bishop of Chester. He died in London on November 19, 1672, reputedly saying on his deathbed that he was "prepared for the great experiment."[11]

SIR CHRISTOPHER WREN (1632–1723), ASTRONOMER AND ARCHITECT

Wren is arguably the most famous of the twelve found-ing members of the Royal Society, although not because of his contributions to experimental philosophy or the mathematical sciences. The architect of St. Paul's Cathe-dral, fifty-six London churches, Chelsea Hospital, and the state apartments at Hampton Court Palace was born into a distinguished family of Anglican clergy: his father was dean of Windsor and his uncle bishop of Ely. Both men were high church Royalists, and both lost their places when the Civil War broke out in 1642. Wren was sent to Westminster School, but around 1647 he contracted an unspecified illness and was placed under the care of the Royalist physician Charles Scarburgh. Scarburgh in-troduced him to a small group of experimental philoso-phers who met every week to discuss "physic, anatomy, geometry, astronomy, navigation, statics . . . and natural experiments." John Wilkins, John Wallis, and Jonathan Goddard were all members of this group. After Wilkins moved to Wadham College, Wren was admitted there to study for his bachelor's degree, and he continued to play an active role in the Oxford Philosophical Club that centered on Wilkins until 1657, when he moved to Gresham to take up the chair of astronomy at the early age of twenty-five. His interests encompassed anatomy, astronomy and lens-grinding, ciphers, fortifications and military engines, a double-writing instrument, a "strainer of the breath, to make the same air serve in respiration," an invention "to weave many ribbons at once with only turning a wheel," a cheap method of embroidering bed-hangings, whale-fishing, water-pumps and "ways of submarine navigation," a weather-clock and weather-wheel, even new musical instruments.[12]

On February 5, 1661, two months after the inaugural meeting of the Society, Wren resigned from Gresham to succeed Seth Ward as Savilian Professor of Astronomy at Oxford. His involvement with the Society was necessarily patchy from then on, and the architectural opportunities presented by the Great Fire of London in 1666 led him further toward the career for which he is famous. He was appointed surveyor of the king's works in 1669, the most important architectural post in the country, but he remained actively involved in the work of the Society, serving as president from 1680 until 1682. He also retained his interest in astronomy and the mathematical sciences: right up to the end of his long life he was still working on astronomical solutions to the problem of longitude.

APPENDIX 2:
THE FIRST CHARTER,
JULY 15, 1662

CHARLES THE SECOND, BY THE GRACE OF GOD KING of England, Scotland, France, and Ireland, Defender of the Faith, &c., to all to whom these present Letters shall come, greeting.

We have long and fully resolved with Ourself to extend not only the boundaries of the Empire, but also the very arts and sciences. Therefore we look with favour upon all forms of learning, but with particular grace we encourage philosophical studies, especially those which by actual experiments attempt either to shape out a new philosophy or to perfect the old. In order, therefore, that such studies, which have not hitherto been sufficiently brilliant in any part of the world, may shine conspicuously amongst our people, and that at length the whole world of letters may always recognize us not only as the Defender of the Faith, but also as the universal lover and patron of every kind of truth:

Know ye that we, of our special grace and of our certain knowledge and mere motion, have ordained, established, granted, and declared, and by these presents for us, our heirs, and successors do ordain, establish, grant, and declare, that from henceforth for ever there shall be a Society, consisting of a President, Council, and Fellows, which shall be called and named The Royal Society; And for us, our heirs, and successors we do make, ordain, create, and constitute by these presents the same Society, by the name of The President, Council, and Fellows of the Royal Society, one body corporate and politic in fact, deed, and name, really and fully, and that by the same name they may have perpetual succession; and that they and their successors (whose studies are to be applied to further promoting by the authority of experiments the sciences of natural things and of useful arts), by the same name of The President, Council, and Fellows of the Royal Society aforesaid, may and shall be in all future times persons able and capable in law to have, acquire, receive, and possess lands and tenements, meadows, feedings, pastures, liberties, privileges, franchises, jurisdictions, and hereditaments whatsoever, to themselves and their successors in fee and perpetuity, or for term of life, lives, or years, or otherwise in whatsoever manner, and also goods and chattels, and all other things, of whatsoever kind, nature, sort, or quality they may be; and also to give, grant, demise, and assign the same lands, tenements, and hereditaments, goods and chattels, and to do and execute all acts and things necessary of and concerning the same, by the name aforesaid; And that by the name of The President, Council, and Fellows of the Royal Society aforesaid they may henceforth for ever be able and have power to plead and be impleaded, to answer and be answered, to defend and be defended, in whatsoever Courts and places, and before whatsoever Judges and Justices and other persons and officers of us, our heirs, and successors, in all and singular actions, pleas, suits, plaints, causes, matters, things, and demands whatsoever, of whatsoever kind, nature, or sort they may or shall be, in the same manner and form as any of our lieges within this our Realm of England, being persons able and capable in law, or as any body corporate or politic within this our Realm of England, may be able and have power to have, acquire, receive, possess, give, and grant, to plead and be impleaded, to answer and be answered, to defend or be defended; And that the same President, Council, and Fellows of the Royal Society aforesaid, and their successors, may have for ever a Common Seal, to serve for transacting the causes and affairs whatsoever of them and their successors; and that it may and shall be good and lawful to the same President, Council, and Fellows of the Royal Society aforesaid, and to their successors for the time being, to break, change, and make anew that Seal from time to time, according as it shall seem most expedient to them.

And that our royal intention may obtain the better effect, and for the good rule and government of the aforesaid Royal Society from time to time, we will, and by these presents for us, our heirs, and successors do grant to the same President, Council, and Fellows of the Royal Society aforesaid, and to their successors, that henceforth for ever the Council aforesaid shall be and consist of twenty-one persons (of whom we will the President to be always one); And that all and singular other persons who within one month next following after the date of these presents shall be received and admitted by the President and Council, and in all time following by the President, Council, and Fellows, into the same Society, as Members of the Royal Society aforesaid, and shall have been noted in the Register by them to be kept, shall be and shall be called and named Fellows of the Royal Society aforesaid: whom, the more eminently they are distinguished for the study of every kind of learning and good letters, the more ardently they desire to promote the honour, studies, and advantage of this Society, the more they are noted for integrity of life, uprightness of character, and piety, and excel in fidelity and affection of mind towards us, our Crown, and dignity, the more we wish them to be especially deemed fitting and worthy of being admitted into the number of the Fellows of the same Society.

And for the better execution of our will and grant in this behalf, we have assigned, nominated, constituted, and made, and by these presents for us, our heirs, and successors do assign, nominate, constitute, and make, our very well-beloved and trusty William, Viscount Brouncker, Chancellor to our very dear consort, Queen Catherine, to be and become the first and present President of the Royal Society aforesaid; willing that the aforesaid William, Viscount Brouncker, shall continue in the office of President of the Royal Society aforesaid from the date of these presents until the feast of St Andrew next following after the date of his presents, and until one other of the Council of the Royal Society aforesaid for the time being shall have been elected, appointed, and sworn to that office in due manner, according to the ordinance and provision below in these presents expressed and declared (if the aforesaid William, Viscount Brouncker, shall live so long); having first taken a corporal oath well and faithfully to execute his office in and by all things touching that office, according to the true intention of these presents, before our very well-beloved and very trusty Cousin and Councillor Edward, Earl of Clarendon, our Chancellor of England: to which same Edward, Earl of Clarendon, our Chancellor aforesaid, we give and grant full power and authority, to administer the oath aforesaid in these words following, that is to say:

I, William, Viscount Brouncker, do promise to deal faithfully and honestly in all things belonging to the trust committed to me as President of this Royal Society, during my employment in that capacity. So help me God!

We have also assigned, constituted, and made, and by these presents for us, our heirs, and successors do make, our beloved and trusty Robert Moray, Knight, one of our Privy Council in our Realm of Scotland; Robert Boyle, Esquire; William Brereton, Esquire, eldest son of the Baron de Brereton; Kenelm Digby, Knight, Chancellor to our very dear mother, Queen Maria; Paul Neile, Knight, one of the Gentlemen of our Privy Chamber; Henry Slingesby, Esquire, another of the Gentlemen of our aforesaid Privy Chamber; William Petty, Knight; John Wallis, Doctor in Divinity;

Timothy Clarke, Doctor in Medicine and one of our Physicians; John Wilkins, Doctor in Divinity; George Ent, Doctor in Medicine; William Aerskine, one of our Cup-bearers; Jonathan Goddard, Doctor in Medicine and Professor of Gresham College; Christopher Wren, Doctor in Medicine, Saville Professor of Astronomy in our University of Oxford; William Balle, Esquire; Matthew Wren, Esquire; John Evelyn, Esquire; Thomas Henshawe, Esquire; Dudley Palmer, of Grey's Inn, in our County of Middlesex, Esquire; and Henry Oldenburg, Esquire, together with the President aforesaid, to be and become the first and present twenty-one of the Council of the Royal Society aforesaid; to be continued in the same offices from the date of these presents until the aforesaid feast of Saint Andrew the Apostle next following, and thenceforth until other fitting and able and sufficient persons shall have been elected, appointed, and sworn into the offices aforesaid (if they shall live so long, or shall not have been amoved [dismissed] for any just and reasonable cause); first taking corporal oaths before the President of the aforesaid Royal Society, well and faithfully to execute their offices in and by all things touching those offices, according to the form and effect of the aforesaid oath, *mutatis mutandis*, to be administered to the President of the Royal Society aforesaid by our Chancellor of England; (to which same President for the time being, for us, our heirs, and successors, we give and grant by these presents full power and authority to administer the oaths aforesaid). And that the same persons, so as it is aforesaid elected, appointed, and sworn, and hereafter to be elected, appointed, and sworn from time to time, to the Council of the aforesaid Royal Society, shall be and become aiding, counselling, and assistant in all matters, business, and affairs touching or concerning the better regulation, government, and direction of the aforesaid Royal Society, and of every Member of the same.

And further we will, and by these presents for us, our heirs, and successors do grant to the aforesaid President, Council, and Fellows of the Royal Society aforesaid, and to their successors, that the President, Council, and Fellows of the Royal Society aforesaid for the time being (of whom we will the President for the time being to be one) may and shall have from time to time in all future times for ever power and authority to nominate and elect, and that they may be able and have power to elect and nominate, every year, on the aforesaid feast of St Andrew, one of the Council of the aforesaid Royal Society for the time being, who may and shall be President of the Royal Society aforesaid until the feast of St Andrew the Apostle thereafter next following (if he shall live so long, or shall not be amoved meanwhile for any just and reasonable cause), and thenceforth until another shall have been elected, appointed, and nominated to the office of President of the Royal Society aforesaid; and that he, after that he shall so have been elected and nominated, as it is aforesaid, to the office of President of the Royal Society aforesaid, before he be admitted to that office, shall take a corporal oath before the Council of the same Royal Society, or any seven or more of them, rightly, well, and faithfully to execute that office in all things touching that office, according to the form and effect of the aforesaid oath, *mutatis mutandis* (to which same Council, or to any seven or more of them, we give and grant and by these presents for us, our heirs, and successors, full power and authority to administer the aforesaid oath from time to time, as often as it shall be necessary); and that after having so taken such oath, as it is aforesaid, he may be able and have power to execute the office of President of the Royal Society aforesaid until the feast of St Andrew the Apostle thereafter next following; And if it shall happen that the President of the Royal Society aforesaid for the time being, at any time, so long as he shall be in the office of President of the same Royal Society, shall die or be amoved from his office, that then and so often it may and shall be good and lawful to the Council and Fellows of the aforesaid Royal Society, or to any seven or more of them (of whom we will the President of the Council aforesaid to be always one at such an election), to elect and appoint another of the aforesaid number of the Council aforesaid as President of the Royal Society aforesaid; and that he so elected and appointed may have and exercise that office during the residue of the same year, and until another shall have been in due manner elected and sworn to that office, first taking a corporal oath in the form last specified; and so as often as the case shall so happen.

And further we will, that whenever it shall happen that any one or any of the Council of the Royal Society aforesaid for the time being shall die, or be amoved from that office, or retire (which same [members] of the Council of the Royal Society aforesaid, and every one of them, we will to be amovable for misbehaviour or any other reasonable cause, at the good pleasure of the President and of the rest of the Council aforesaid surviving and remaining in that office, or of the major part of the same, of whom we will the President of the Royal Society aforesaid for the time being to be one), to nominate, elect, and appoint one other or several others of the Fellows of the Royal Society aforesaid, in the place or places of him or them so dead, retired, or amoved, to fill up the aforesaid number of twenty-one persons of the Council of the Royal Society aforesaid; and that he or they so elected and appointed in that office may have the same office until the feast of St Andrew the Apostle then next following, and thenceforth until one other or several others shall have been elected, appointed, and nominated; first taking a corporal oath before the President and Council of the Royal Society aforesaid, or any seven or more of them for the time being, well and faithfully to execute that office in and by all things touching that office, according to the true intention of these presents.

And further we will, and by these presents for us, our heirs, and successors do grant to the aforesaid President, Council, and Fellows of the aforesaid Royal Society, and to their successors, that they and their successors, every year, on the aforesaid feast of St Andrew the Apostle, may and shall have full power and authority to elect, nominate, appoint, and change ten of the Fellows of the Royal Society aforesaid, to fill up the places and offices of ten of the aforesaid number of twenty-one of the Council of the Royal Society aforesaid; for we do declare it to be our royal pleasure, and by these presents for us, our heirs, and successors we do grant, that ten of the aforesaid Council, and no more, shall be annually changed and amoved by the President, Council, and Fellows of the Royal Society aforesaid.

We will also, and for us, our heirs, and successors do grant to the aforesaid President, Council, and Fellows of the aforesaid Royal Society, and to their successors, that if it shall happen that the President of the same Royal Society for the time being is detained by sickness or infirmity, or is employed in the service of us, our heirs, or successors, or is otherwise occupied, so that he shall not be able to attend to the necessary affairs of the same Royal Society touching the office of President, that then and so often it may and shall be good and lawful to the same President so detained, employed, or occupied, to nominate and appoint one of the Council of the aforesaid Royal Society for the time being to be and become the Deputy of the same President; which same Deputy, so to be made and appointed in the office of Deputy of the President aforesaid, may and shall be the Deputy of the same President from time to time, as often as the aforesaid President shall happen to be so absent, during the whole time in which the aforesaid President shall continue in the office of President; unless in the meanwhile the aforesaid President of the Royal Society aforesaid for the time being shall have made and appointed one other of the aforesaid Council his Deputy; And that every such Deputy of the aforesaid President so to be made and appointed, as it is aforesaid, may be able and have power to do and execute all and singular things which pertain or ought to pertain to the office of President of the aforesaid Royal Society, or which are limited and appointed to be done and executed by the aforesaid President, by virtue of these our Letters Patent, from time to time, as often as the aforesaid President shall so happen to be absent, during such time as he shall continue the Deputy of the aforesaid President, by force of these our Letters Patent, as fully, freely, and wholly, and in as ample manner and form, as the aforesaid President, if he were present, would be able and have power and do and execute those things; a corporal oath first to be taken by such Deputy upon the holy Gospels of God, in the form and effect last specified, well and faithfully to execute all and singular things which pertain to the office of President, before the aforesaid Council of the aforesaid Royal Society, or any seven or more of them; and so as often as the case shall so happen; to which same Council, or to any seven or more of them, for the time being, we do give and grant by these presents power and authority to administer the oath aforesaid as often as the case shall so happen, without procuring or obtaining any writ, commission, or further warrant in that behalf from us, our heirs or successors.

And further we will, and by these presents for us, our heirs, and successors do grant to the aforesaid President, Council, and Fellows of the Royal Society aforesaid, and to their successors, that they and their successors henceforth for ever may and shall have one Treasurer, two Secretaries, one Clerk, and two Serjeants-at-Mace, who may from time to time attend upon the President; and that the aforesaid Treasurer, Secretaries, Clerk, and Serjeants-at-Mace, to be elected and nominated, before they be admitted to execute their several and respective offices, shall take their corporal oaths in the form and effect last specified, before the President and Council of the same Royal Society, or any seven or more of them, rightly, well, and faithfully to execute their several and respective offices in all things touching the same; and that after having so taken such oaths, as it is aforesaid, they may exercise and use their respective offices; to which same President and Council, or to any seven or more of them, we do give and grant by these presents full power and authority to administer the oaths aforesaid from time to time to the aforesaid several and respective officers and their successors: And we have assigned, nominated, chosen, created, appointed, and made, and by these presents for us, our heirs, and successors do assign, nominate, choose, create, appoint, and make, our beloved subjects William Balle, Esquire, to be and become the first and present Treasurer, and the aforesaid John Wilkins and Henry Oldenburg to be and become the first and present Secretaries, of the aforesaid Royal Society; to be continued in the same offices until the aforesaid feast of St Andrew the Apostle next following after the date of these presents: And that from time to time and at all times on the aforesaid feast of Saint Andrew the Apostle (unless it shall be Sunday, and if it be Sunday, then on the day next following) the President, Council, and Fellows of the aforesaid Royal Society for the time being, or the major part of the same (of whom we will the President for the time being to be one), may be able and have power to elect, nominate, and appoint other upright and discreet men, from time to time, as Treasurer, Secretaries, Clerk, and Sergeants-at-Mace of the aforesaid Royal Society; and that those who shall so have been elected, appointed, and sworn to the aforesaid several and respective offices, as it is aforesaid, may be able and have power to exercise and enjoy those respective offices until the aforesaid feast of St Andrew then next following, their aforesaid oaths, as it is aforesaid, first to be taken; and so as often as the case shall so happen: And if it shall happen that any one or any of the officers aforesaid of the same Royal Society shall die, or be amoved from their respective offices, that then and so often it may and shall be good and lawful to the President, Council, and Fellows of the aforesaid Royal Society, or to the major part of the same (of whom we will the President for the time being to be one), to elect and appoint another or others to the office or offices of those persons so deceased or amoved; and that he or they so elected and appointed may have and exercise the respective offices aforesaid during the residue of the same year, and until another or others shall have been in due manner elected and sworn to those respective offices; and so as often as the case shall so happen.

And moreover we will, and of our special grace and of our certain knowledge and mere motion do grant to the aforesaid President, Council, and Fellows of the Royal Society aforesaid, and to their successors, that the President and Council of the aforesaid Royal Society for the time being, and the major part of the same (of whom we will the President for the time being to be one), may be able and have power to meet together and assemble in a College or other public place or Hall within our City of London, or in any other convenient place within ten miles of our same City; and that they so met together and assembled shall and may have full authority, power, and faculty from time to time to draw up, constitute, ordain, make, and establish such laws, statutes, acts, ordinances, and constitutions as shall seem to them, or to the major part of them, to be good, wholesome, useful, honourable, and necessary, according to their sound discretions, and to do and perform all other things whatsoever belonging to the affairs and matters of the Royal Society aforesaid; all and singular which laws, statutes, acts, ordinances, and constitutions so to be made as it is aforesaid, we will, and by these presents for us, our heirs, and successors, firmly enjoining, do order and command, that they shall be inviolably observed from time to time, according to the tenor and effect of the same: so nevertheless that the aforesaid laws, statutes, acts, ordinances, and constitutions so to be made as it is aforesaid, and every one of them, be reasonable, and not repugnant or contrary to the laws, customs, acts, or statutes of this our Realm of England.

And further, of our more ample special grace and of our certain knowledge and mere motion, we have given and granted, and by these presents for us, our heirs, and successors do give and grant to the aforesaid Council and Fellows of the aforesaid Royal Society, and to their successors, full power and authority from time to time, to elect, nominate, and appoint one or more Typographers or Printers, and Chalcographers or Engravers, and to grant to him or them, by a writing, sealed with the Common Seal of the aforesaid Royal Society, and signed by the hand of the President for the time being, faculty to print such things, matters, and affairs touching or concerning the aforesaid [Royal] Society, as shall have been committed to the aforesaid Typographer or Printer, Chalcographer or Engraver, or Typographers or Printers, Chalcographers or Engravers, from time to time, by the President and Council of the aforesaid Royal Society, or any seven or more of them (of whom we will the President for the time being to be one); their corporal oaths first to be taken, before they be admitted to exercise their offices, before the President and Council for the time being, or any seven or more of them, in the form and effect last specified; to which same President and Council, or to any seven or more of them, we do give and grant by these presents full power and authority to administer the oaths aforesaid.

And further, in order that the aforesaid President, Council, and Fellows of the aforesaid Royal Society may obtain the better effect in their philosophical studies, of our more ample special grace and of our certain knowledge and mere motion, we have given and granted, and by these presents for us, our heirs, and successors, do give and grant, to the aforesaid President, Council, and Fellows of the aforesaid Royal Society, and to their successors, that they and their successors from time to time may and shall have full power and authority from time to time, and at such seasonable times, according to their discretion, to require, take, and receive the bodies of such persons as have suffered death by the hand of the executioner, and to anatomize them, in such ample form and manner, and to all intents and purposes, as the College of Physicians and the Corporation of Surgeons of our City of London have used or enjoyed, or may be able and have power to use or enjoy, the same bodies.

And further, for the improvement of the experiments, arts, and sciences of the aforesaid Royal Society, of our more abundant special grace and of our certain knowledge and mere motion, we have given and granted, and by these presents for us, our heirs and successors do give and grant, to the aforesaid President, Council, and Fellows of the aforesaid Royal Society, and to their successors, that they and their successors from time to time may and shall have full power and authority, by letters or epistles under the hand of the aforesaid President, in the presence of the Council, or of any seven or more of them, and in the name of the Royal Society, and sealed with their Common Seal aforesaid, to enjoy mutual intelligence and knowledge with all and all manner of strangers and foreigners, whether private or collegiate, corporate or politic, without any molestation, interruption, or disturbance whatsoever: Provided nevertheless, that this our indulgence, so granted as it is aforesaid, be not extended to further use than the particular benefit and interest of the aforesaid Royal Society in matters or things philosophical, mathematical, or mechanical.

And further we have given and granted, and by these presents for us, our heirs, and successors do give and grant to the aforesaid President, Council, and Fellows of the Royal Society aforesaid, and to their successors, full power and authority to erect, build, and construct, or to make or cause to be erected, built, and constructed, within our City of London, or ten miles of the same, one or more College or Colleges, of whatsoever kind or quality, for the habitation, assembly, and meeting of the aforesaid President, Council, and Fellows of the aforesaid Royal Society, and of their successors, for the ordering and arranging of their affairs and other matters concerning the same Royal Society.

And further we will, and by these presents for us, our heirs, and successors do ordain, constitute, and appoint, that if any abuses or differences hereafter shall arise and happen concerning the government or other matters or affairs of the aforesaid Royal Society, whereby any injury or hindrance may be done to the constitution, stability, and progress of the studies, or to the matters and affairs, of the same; that then and so often, by these presents, for us, our heirs, and successors, we do authorize, [ordain], nominate, assign, and appoint our aforesaid very well-beloved and very trusty Cousin and Councillor Edward, Earl of Clarendon, our Chancellor of our Realm of England, by himself during his life, and, after his death, then the Archbishop of Canterbury, the Chancellor or Keeper of the Great Seal of England, the Treasurer of England, the Bishop of London, the Keeper of the Privy Seal, and the two Principal Secretaries for the time being, or any four or more of them, to reconcile, compose, and adjust the same differences and abuses.

And further we will, and by these presents, for us, our heirs, and successors, firmly enjoining, do order and command all and singular the Justices, Mayors, Aldermen, Sheriffs, Bailiffs, Constables, and other officers, ministers, and subjects whomsoever of us, our heirs, and successors, that they be from time to time aiding and assistant to the aforesaid President, Council, and Fellows of the Royal Society aforesaid, and to their successors, in and by all things, according to the true intention of these our Letters Patent.

Although express mention of the true yearly value or of the certainty of the premises, or of any of them, or of other gifts or grants before these times made by us or by any of our progenitors or predecessors to the aforesaid President, Council, and Fellows of the Royal Society, is not made in these presents; or any statute, act, ordinance, provision, proclamation, or restriction to the contrary thereof heretofore had, made, enacted, ordained, or provided, or any other thing, cause, or matter whatsoever, in any wise notwithstanding.

In witness whereof we have caused these our Letters to be made Patent. Witness Ourself, at Westminster, the fifteenth day of July, in the fourteenth year of our reign.

By the King himself.
HOWARD.

APPENDIX 3:
EXPERIMENTS AND
OBSERVATIONS

I

An Account of the Experiment of Transfusion, *practised upon a* Man in London.

This was perform'd, Novemb. 23. 1667. *Upon one Mr.* Arthur Coga, *at* Arundel-*House, in the presence of many considerable and intelligent persons, by the management of those two Learned Physitians and dextrous Anatomists* Dr. Richard Lower, *and Dr.* Edmund King, *the latter of whom communicated the Relation of it, as followeth.*

The Experiment of Transfusion of Blood into an *humane* Veine was made by Us in this manner. Having prepared the *Carotid* Artery in a young Sheep, we inserted a Silver-Pipe into the Quills to let the Blood run through it into a Poringer, and in the space of almost a *minut*, about 12. ounces of the Sheeps-bloud ran through the Pipe into the Poringer; which was somewhat to direct us in the quantity of Bloud now to be Transfus'd into the Man. Which done, when we came to prepare the *Veine* in the *Man's* Arme, the Veine seem'd too small for that Pipe, which we intended to insert into it; so that we imployed another, about one third part lesse, at the little end. Then we made an incision in the Veine, after the Method, formerly publisht *Numb*. 28; which Method we observ'd without any other alteration, but in the shape of one of our Pipes; which we found more convenient for our purpose. And, having open'd the Veine in the Man's Arme, with as much ease as in the common way of Venæ-section, we let thence run out 6 or 7 ounces of Blood. Then we planted our silver-pipe into the said Incision, and inserted Quils between the two Pipes already advanced in the two subjects, to convey the *Arteriall* bloud from the Sheep into the Veine of the Man. But this Blood was near a *minut*, before it had past through the Pipes and Quills into the Arme; and then it ran freely into the Mans veine for the space of 2. *minutes* at least; so that we could feel a *Pulse* in the said veine just beyond the end of the Silver-pipe; though the Patient said, he did not feel the blood *hot* (as was reported of the subject in the *French* Experiment) which may very well be imputed to the length of the Pipes, through which the blood passed, losing thereby so much of its Heat, as to come in a temper very agreeable to Venal Blood. And as to the quantity of Blood receiv'd into the Man's Veine, we Judge, there was about 9. or 10. ounces: For, allowing this pipe ⅓ lesse, than that, through which 12. ounces pass'd in *one* minute before, we may very well suppose, it might in *two* minuts conveigh as much blood into the Veine, as the other did into the Porringer, in *one* minut; granting withall, that the Bloud did not run so vigorously the second minut, as it did the first, nor the third, as the second, *&c.* But, that the Blood did run all the time of those two minutes, we conclude from thence; *First*, because we felt a Pulse during that time; *Secondly*, because when upon the Man's saying, He thought, he had enough, we drew the pipe out of his Veine, the Sheeps-bloud ran through it with a full stream; which it had not done, if there had been any stop before, in the space of those two minutes; the bloud being so very apt to coagulate in the Pipes upon the least stop, especially the Pipes being so long as three Quills.

The Man *after* this operation, as well as *in* it, found himself very well, and hath given in his own Narrative under his own hand, enlarging more upon the benefit, he thinks, he hath received by it, than we think fit to own as yet. He urged us to have the Experiment repeated upon him within three or four dayes after this; but it was thought advisable, to put it off somewhat longer. And the next time, we hope to be more exact, especially in weighing the Emittent Animal before and after the Operation, to have a more Just account of the quantity of Bloud, it shall have lost.

[*Philosophical Transactions* 2, no. 30 (1667). Edmund King made a second experiment of transfusing sheep's blood into Arthur Coga in December 1667. Coga appeared at a meeting of the Royal Society on December 19 to say that "he found himself very well at present, though he had been at first somewhat feverish upon it; which was imputed to his excess in drinking too much wine soon after the operation" (Birch, *History of the Royal Society*, vol. 2, 227). Enthusiasm for transfusion experiments waned after 1668 when a French surgeon, Jean-Baptiste Denis, who had been pursuing a similar line of research to Lower and King, killed a human subject.]

II

Some Observations about Shining Flesh, *made by the Honourable* Robert Boyle; *Febr. 15. 1671/72; and by way of Letter addressed to the Publisher, and presented to the R. Society.*

Yesternight when I was about to go to bed, an *Amanuensis* of mine, accustom'd to make Observations, informed me, that one of the Servants of the house, going upon some occasion into the Larder, was frighted by something of Luminous that she saw (notwithstanding the darkness of the place,) where the meat had been hung up before: Whereupon suspending for a while my going to rest, I presently sent for the meat into my Chamber, and caused it to be placed in a corner of the room capable of being made considerably dark, and then I plainly saw, both with wonder and delight, that the joint of meat did in divers places shine like rotten Wood or stinking Fish; which was so uncommon a sight, that I had presently thoughts of inviting you to be a sharer in the pleasure of it. But the late hour of the night did not only make me fear to give you too unseasonable a trouble, but being joyned with a great Cold I had got that day by making Tryal of a new Telescope (you saw,) in a windy place, I durst not sit up long enough to make all the tryals that I thought of and judg'd the occasion worthy of. But yet, because I effectually resolved to imploy the little time I had to spare, in making such Observations and tryals, as the accommodations, I could procure at so inconvenient an hour, would enable me, I shall here give you a brief account of the chief circumstances and *Phœnomena*, that I had opportunity to take notice of.

1. Then I must tell you, that the subject, we discourse of, was a Neck of Veal, which, as I learned by inquiry, had been bought of a Country-butcher on the Tuesday preceding.

2. In this one piece of meat I reckoned distinctly above *twenty* several places that did all of them shine, though not all of them alike, some of them doing it but very faintly.

3. The bigness of these Lucid parts was differing enough, some of them being as big as the nail of a man's middle finger, some few bigger, and most of them less. Nor were there figures at all more uniform, some being inclined to a round, others almost oval, but the greatest part of them very irregularly shap'd.

4. The parts that shone most, which 'twas not so easie to determine in the dark, were some gristly or soft parts of the bones, where the Butcher's Cleaver had passed; but these were not the only parts that were luminous; for by drawing to and fro the *Medulla spinalis*, we found, that a part of that also did not shine ill: And I perceived one place in a *Tendon* to afford some light; and lastly three or four spots in the fleshy parts at a good distance from the bones were plainly discovered by their own light, though that were fainter than in the parts above mentioned.

5. When all these Lucid parts were survey'd together, they made a very splendid shew; but 'twas not so

easie, because of the moistness and grossness of the lump of matter, to examine the degree of their Luminousness, as it is to estimate that of Gloworms, which being small and dry bodies may be conveniently laid in a book, and made to move from one letter or word to another. But by good fortune having by me the curious *Transactions* of this month, I was able so to supply that flexible paper to some of the more resplendent spots, that I could plainly read divers consecutive letters of the Title.

[...]

18. What further *Phœnomena* this morning might have afforded me, I cannot tell, having been hastily called up before day for a Niece, that I am very justly and exceedingly concerned for, who was thought to be upon the point of death, and whose almost gasping condition had too much affected and imployed me, to leave me any time for Philosophical entertainments, that require a calm, if not a pleased, mind. Only this I took notice of, because the observation could not cost me a minute of an hour, that whilst they were bringing me Candles for to rise by, I looked upon a clean phial, that I had laid upon the bed by me after a piece of our luminous Veal had been included in it, and found it to shine vividly at that time, which was between four and five of clock this morning, since when I have made no one observation or tryal.

[...]

[*Philosophical Transactions* 7, no. 89 (1672). Boyle's niece recovered, and he invited Henry Oldenburg—to whom this letter was addressed—and several other members of the Royal Society to his house to view the shining veal. It still glowed six days after the initial discovery, although by the seventh day the luminescence had completed subsided.]

III

Part of two Letters from Mr **Anthony Van Leewenhoek,** *F.R.S. concerning Worms pretended to be taken from the Teeth.*

Delph in *Holland, July* 27, 1700.

SIR,

Your most acceptable Letter, dated *July* 4. old stile, came to hand on the 19th of the same month, new stile: upon the receipt whereof, I immediately open'd the piece of black Silk contain'd within it, where I found two little Worms dead, and one alive, which were sent you to be convey'd to me, as being taken out of a corrupt Tooth by smoaking.

Immediately I made it my business to find out what sort of Worms they were, and how they were generated.

I had not spent much time upon the living Worm, which wanted above one half of its full growth, when I concluded that it sprung from the Egg of a small Fly, of that sort of Flies that frequent Cheesemongers Shops for the most part, and have a peculiar humour of laying their Eggs upon Cheese: now the Worms taking their rise from those Eggs, bore through the Cheese, and take their nourishment and growth from it; and afterwards become little Flies again. When these little Worms arrive at such a bulk as is discernible by the naked eye, we call 'em Worms.

I took a Glass Tube, one end whereof was blown together, it being somewhat more than a finger long, and about half an Inch wide; and put into it the living Worm, together with some crums of very old and fat Cheese, in order to try if the Worm feeding upon the Cheese would come to its full growth.

I stopp'd the Glass with a Cork; for I am positive that a Worm may live and grow in a Glass as well as in a firm Cheese cover'd up all about.

Being confident that both the dead Worms and the living one were of the above-mention'd sort, I got a Cheesemonger to single out that sort of old Cheese, and bring some of its little Worms to my house.

I put 6 or 8 of the greatest of these Worms in two distinct little Glasses, together with one of the dead Worms you sent me, designing to compare the living and dead Worms before a Magnifying-Glass: and could not descry the least difference either in the Head or the whole Body.

When these Worms had been shut up 5 days without any food, I observ'd 'em gnawing the Corks that stopp'd the Glasses. Then I put in a little Cheese, that if they did not arrive at their full Growth, they might not want Food, in order to their change into Flies.

I likewise endeavour'd to bring one of these Worms to an extended and quiet posture, in order to view the internal parts; which succeeded with me several times: and so I saw to my great admiration such moving instruments all over the Body within, that not one of a thousand would be perswaded to believe, that in such a contemptible Insect there is so much to be seen: for in one place I thought I saw the motion of the Heart, and not far from that the motion of the Stomach: But after all the narrowest enquiry, I could not descry any motion in the Blood in those parts which I took for Veins. If one of these Worms were cut up, and its Bowels presented to view, they'll present you with an astonishing number of Vessels, which appear'd to me like Blood Vessels.

I would fain know what food these Creatures live upon, besides Cheese; in the first place, because I never yet saw 'em any where else but in Cheese; and in the next place, because they cannot arise from Flesh; for the Worms that spring from the Eggs of Flies in Flesh, which we also call Mites or Fly-bows, are fully grown in the space of nine days; but those which feed upon Cheese require a longer time for com-pleating their growth, and Flesh will not keep so long without Salting or Smoaking.

Since things stand thus, we must conclude that it is always natural to these little Flies to lay their Eggs upon such substances as will not easily cor-rupt; now we may justly reckon Cheese to be such.

Let's imagine that the Patient, whose Tooth threw out the Worms by Smoaking, had some time before eaten Cheese laden with young Worms, or Eggs of the above-mention'd Flies, and that these Worms or Eggs were not touch'd or injur'd in the chewing of the Cheese, but stuck in the hollow Teeth, and at last insinu-ated themselves so far into the substance of the Teeth, that they gnaw'd the sensible parts, and so occasion'd the great pain.

It appears very strange to me, that smoak in the Mouth or Tooth should have such an effect, as to bring Worms out of hollow Teeth; for I cannot conceive how the little Worms should have a respiration, to be so far prejudic'd by the Smoak, that they are obliged to come forth.

To satisfy my self upon this point, I took a Glass Ball, the Diameter whereof was almost three inches, with a little hole in it as large as the thickness of a Goose Quill. In this Glass I put then Cheese-worms of the largest sort, and twice or thrice threw in burning Brimstone, to the breadth of the back of a Knife, upon a thin slender piece of Hemp; and observed that the burning Brimstone did not at all injure the Worms, so far as I could see; and about an hour after the burning of the Sulphur, I put the mouth of the Glass to my Nose, and could still plainly perceive the smell of the Sulphur.

'Tis well known that when we burn Sulphur, the Sulphur does not perish, but only is divided into such small particles, as are remov'd from our view: and accordingly in this experiment I saw an infinite number of very small sulphureous particles, sticking to the inside of the Glass, which appeared to me to be round.

I have often fill'd such a Glass Ball with Water, and exposed it to the rays of the Sun, (instead of a grounded Glass) which were so concentred that it burnt Paper.

I return thanks for the communication of these Worms; especially, because heretofore several people wanted to know of me from whence they came, and added, that they did not come by generation, but started up of themselves; the contrary of which must be looked upon as a certain truth.

The living Worm that you was so kind as to send me, I have kept still alive, and cannot but think that 'tis become bigger since I had it. I'll try to bring it up so far, till it turns into a Fly.

These Worms have a hard and strong skin, and may last a long time.

I remember, some years ago my late Wife being much afflicted with the Tooth-ach, complain'd that the pain was such, as if the Flesh had been gnaw'd through.

Having called a Doctor of Physick oftner than once, a great many Remedies were try'd to no purpose: at last she found benefit by dropping the Oyl of Vitriol into the hollow Tooth, which

I did with a Glass Instrument, that convey'd the Oyl to the Tooth without injuring the Muscles.

Now it's possible she might have got one or more of these little Worms in the hollow Tooth, at a time when she eat heartily of old Cheese, which was seiz'd with whitish rotten-ness, and had a great many little Worms in it, which she did not observe, tho she fed often upon it. Upon this supposition the pain might be occasion'd by those Worms, which were afterwards kill'd by the Oyl of Vitriol when we knew nothing of 'em.

[*Philosophical Transactions* 22, no. 265 (1701). The Dutchman Antonie van Leeuwenhoek, often called "the father of microbiology," corresponded regularly with the Royal Society. He was elected a Fellow in 1680. The Leeuwenhoek Medal and Lecture, awarded biennially, commemorates his association with the Society.

NOTES

Full references for all works cited may be found in the Bibliography.

Prelude

1. Joshua 10:13 (KJV); Martin Luther, *Tablebook (Tischreden)*, quoted at Richard Pogge, "A Brief Note on Religious Objections to Copernicus," Department of Astronomy, Ohio State University, www.astronomy.ohio-state.edu/~pogge/Ast161/Unit3/response.html; Selderhuis, *The Calvin Handbook*, 452.

2. B. A., *Sick-man's Rare Jewel*, 14.

3. Ibid., 13.

4. University of Oxford, *Statutes*, vol. 1, 31.

5. Carlo, *Sidereal Messenger*, 10.

6. Ibid., 42–43.

7. Waters, *Art of Navigation*, 299.

8. Bacon, *Francisci de Verulamio*, 19.

9. Clark, "Brief Lives," vol. 1, 75.

10. Francis Bacon, *The New Atlantis*, full text available at Gutenberg, https://www.gutenberg.org/files/2434/2434-h/2434-h.htm.

Chapter 1 Foundation

1. *Newes from the Dead, or a True and Exact Narration of the Miraculous Deliverance of Anne Greene.*

2. Birch, *History of the Royal Society*, vol. 1, 3.

3. Ibid., 4.

4. Pope, *Life of Seth*, 20–21.

5. Wallis, *Defence of the Royal Society*, 7.

6. Ibid., 8.

7. Pope, *Life of Seth*, 29.

8. Robinson, "Unpublished Letter of Dr Seth Ward," 69, 70.

9. Turnbull, "Samuel Hartlib's Influence," 114.

10. Evelyn, *Diary*, vol. 1, 295.

11. Clark, "Brief Lives," vol. 1, 276.

12. J. Ward, *Lives of the Professors*, 241.

13. Copeman, "Dr Jonathan Goddard," 72.

Chapter 2 Charter

1. Birch, *History of the Royal Society*, vol. 1, 4.

2. Ibid., 5.

3. Ibid.

4. Hunter, "Social Basis," *14*.

5. Birch, History of the Royal Society, vol. 1, 7.

6. Ibid., 6.

7. Lyons, *Royal Society*, 27.

8. Evelyn, *Diary*, vol. 1, 365.

9. Birch, *History of the Royal Society*, vol. 1, 50.

10. Ibid., 85.

11. Ibid., 107.

12. Ibid., 104.

13. Royal Society of London, *First Charter* (see p. 151).

14. Birch, *History of the Royal Society*, vol. 4, 144.

Chapter 3 Experiment

1. Birch, *History of the Royal Society*, vol. 4, 138.

2. Sprat, *History of the Royal-Society*, 61.

3. Hunter, *Establishing the New Science*, 223, 224.

4. Birch, *History of the Royal Society*, vol. 1, 5.

5. Ibid., 7.

6. Ibid., 9–10.

7. Ibid., 8.

8. Ibid., 29, 31.

9. Ibid., 17.

10. Ibid., vol. 4, 101.

11. Ibid., vol. 1, 83.

12. Ibid., 66.

13. Ibid., 35.

14. Ibid., 10.

15. Ibid., 288–291.

16. Hall, *Promoting Experimental Learning*, 31.

17. Birch, *History of the Royal Society*, vol. 1, 124.

18. J. Ward, *Lives of the Professors*, 187.

19. Birch, *History of the Royal Society*, vol. 1, 125.

20. Ibid., 179.

21. Hooke, *Micrographia*, dedication.

22. Oldenburg, *Correspondence*, vol. 3, 230–231.

23. Birch, *History of the Royal Society*, vol. 2, 214–215.

24. *Philosophical Transactions*, December 9, 1667, 1.

25. Ibid., 2.

26. Birch, *History of the Royal Society*, vol. 2, 216.

27. 'Espinasse, *Robert Hooke*, 52.

28. Birch, *History of the Royal Society*, vol. 4, 518.

CHAPTER 4 PHILOSOPHICAL TRANSACTIONS

1. *Philosophical Transactions*, vol. 1, no. 1 (March 6, 1665): title page.

2. Ibid., vol. 1, no. 1 (March 6, 1665): 15.

3. Andrade, "Birth and Early Days," 26–27.

4. Royal Society of London, *First Charter* (see p. 151).

5. Kronick, "Notes on the Printing History," 244.

6. Ibid., 245.

7. *Philosophical Transactions*, vol. 1, no. 1 (March 6, 1665): 2.

8. Ibid., vol. 3, no. 44 (January 1, 1668): no pagination.

9. *Philosophical Collections*, no. 1 (1679): 1.

10. Ibid., no. 2 (1681): 10.

11. *Philosophical Transactions*, vol. 13, no. 143 (1683): 2.

12. Ibid.

13. "About Royal Society Open Science," *Royal Society Open Science*, Royal Society Publishing, http://rsos.royalsociety publishing.org/about.

CHAPTER 5 REPOSITORIES AND LABORATORIES

1. Sprat, *History of the Royal-Society*, 434.

2. Royal Society MS ED W 3, 7: Wren to Oldenburg, June 7, 1668.

3. Martin, "Former Homes," 14.

4. Ibid.

5. Macky, *Journey Through England*, 165.

6. Martin, "Former Homes," 16.

7. Sprat, *History of the Royal-Society*, 251.

8. Birch, *History of the Royal Society*, vol. 1, 321.

9. Hubert, *Catalogue of Many Natural Rarities*, title page.

10. Ibid., 1.

11. Ibid., 10.

12. Ibid., 50.

13. Ibid., 66, 67.

14. Hunter, *Establishing the New Science*, 136.

15. Grew, *Musæum Regalis Societatis*, title page.

16. Ibid., 83.

17. Ibid., 2.

18. Thomas, "A 'Philosophical Storehouse,'" 24.

19. Hatton, *New View of London*, vol. 2, 666.

20. *British Curiosities*, 44.

21. E. Ward, *London Spy Compleat*, 59.

22. Thomas, "'Philosophical Storehouse,'" 26.

23. Ibid., 32.

24. *Country Spy*, 43.

25. Thomas, "'Philosophical Storehouse,'" 38.

26. *Royal Society Original Journal Book,* vol. 25, November 17, 1763, 138.

27. Thomas, "'Philosophical Storehouse,'" 121.

Chapter 6 Persons of Great Quality

1. Sorbière, *Voyage to England*, 36.

2. Ibid.

3. Ibid., 37.

4. Evelyn, *Diary*, vol. 2, 154.

5. Birch, *History of the Royal Society*, vol. 4, 92.

6. Ibid., 158.

7. Andrade, *Brief History*, 6.

8. Birch, *History of the Royal Society*, vol. 3, 269.

9. [Thomas Birch], "Memoirs Relating to the Life of Sir Hans Sloane, Formerly President of the Royal Society," British Library, Manuscripts Collection, Add. MS 4241.

10. Andrade, *Brief History*, 8, where the lines are mistakenly said to refer to one of Pringle's predecessors as president, the antiquarian Martin Folkes.

11. Lyons, *Royal Society*, 218.

Chapter 7 Sticks and Stones

1. Evelyn, *Silva*, "To the Reader," no pagination.
2. Ibid.
3. Ibid.
4. Sprat, *History of the Royal-Society*, "Epistle Dedicatory."
5. Ibid., 249.
6. Ibid., 438.
7. South, *Sermons*, vol. 1, 220–221.
8. Ibid., 222.
9. Evelyn, *Diary*, vol. 2, 43.
10. South, *Discourses*, 323.
11. Glanvill, *Plus Ultra*, 7.
12. Sprat, *History of the Royal-Society*, 417.
13. Pepys, *Diary*, vol. 5, 33.
14. Shadwell, *Virtuoso*, Act II, scene 2, ll. 190–194.
15. Ibid., Act V, scene 2, ll. 82–88.
16. Hooke, *Diary*, June 2, 1676.
17. Wotton, *Reflections upon Ancient and Modern Learning*, 393–394.
18. Miller, "Henry Fielding's Satire," 73.
19. Hill, *Dissertation*, 35.
20. Ibid., 24.
21. Ibid., 32–33.
22. Hill, *Review of the Works*, 18, 42, 95.

Chapter 8 Reform

1. Foster, "Note on the History," 509.
2. Granville, *Science Without a Head*, 81.
3. Ibid., 82.
4. Babbage, *Reflections on the Decline*, 53.
5. Ibid., 141.
6. South, *Charges Against the President*, 13.

7. Granville, *Science Without a Head*, 37.

8. Ibid., 49.

9. Ibid., 84.

10. *Times*, October 27, 1830, 5.

11. Ibid., October 29, 1830, 2.

12. Ibid., November 25, 1830, 2.

13. Ibid., December 1, 1830, 2.

14. Lyons, *Royal Society*, 272.

CHAPTER 9 FOREIGN PARTS

1. Birch, *History of the Royal Society*, vol. 2, 418–420.

2. Ibid., vol. 3, 46.

3. Carter, "Royal Society," 246.

4. Ibid.

5. Hornsby, "On the Transit of Venus," 344.

6. Carter, "Royal Society," 249, 251.

7. Ibid., 251.

CHAPTER 10 A BRAVE NEW WORLD

1. Mason, "Women Fellows' Jubilee," 126.

2. Ibid.

3. *Times*, June 22, 1899, 12.

4. Mason, "Hertha Ayrton," 213.

5. Ibid., 211.

6. "Sex Disqualification (Removal) Act 1919," National Archives, United Kingdom, www.legislation.gov.uk/ukpga /Geo5/9-10/71/section/1/enacted.

7. "History of NPL," National Physical Laboratory, www .npl.co.uk/about/history.

8. Hughes, "Divine Right or Democracy?," S115, n. 14.

9. Ibid., 104.

10. Ibid., 102.

11. Ibid., 108.

12. *Manchester Guardian*, December 2, 1935, 12.

13. Ibid.

14. "Anniversary Address from Sir Venki Ramakrishnan, President of the Royal Society," Royal Society, November 30, 2017, https://royalsociety.org/news/2017/11/president -anniversary-address.

15. Royal Society, https://royalsociety.org.

16. "Mission and Priorities," Royal Society, https://royal society.org/about-us/mission-priorities.

Appendices

1. Hooke, *Diary*, 321.

2. Youngson, "Alexander Bruce," 257.

3. Birch, *History of the Royal Society*, vol. 1, 15, 8.

4. Clark, "Brief Lives," vol. 2, 82.

5. Ibid., 144.

6. Pepys, *Diary*, vol. 5, 27.

7. Clark, "Brief Lives," vol. 2, 144.

8. Birch, *History of the Royal Society*, vol. 1, 98.

9. Clark, "Brief Lives," vol. 2, 301.

10. Pope, *Life of Seth*, 29.

11. Henry, "Wilkins, John."

12. Wren, *Parentalia*, 198–199.

BIBLIOGRAPHY

Andrade, E. N. da C. *A Brief History of the Royal Society*. London, 1960.

———. "The Birth and Early Days of the *Philosophical Transactions*." *Notes and Records of the Royal Society of London* 20, no. 1 (June 1965): 9–27.

B. A. *The Sick-man's Rare Jewel*. London, 1674.

Babbage, Charles. *Reflections on the Decline of Science in England and on Some of Its Causes*. London, 1830.

Bacon, Francis. *Francisci de Verulamio, Summi Angliæ Cancellarii, Instauratio Magna*. London, 1620.

Bennett, J. A. "Wren's Last Building?" *Notes and Records of the Royal Society of London* 27, no. 1 (August 1972): 107–118.

Birch, Thomas. *The History of the Royal Society of London*. 4 vols. London, 1756–1757.

British Curiosities in Nature and Art. London, 1713.

Carlo, E. S., ed. and trans. *The Sidereal Messenger of Galileo Galilei and a Part of the Preface to Kepler's Dioptrics*. London, 1880.

Bibliography

Carter, Harold B. "The Royal Society and the Voyage of HMS 'Endeavour,' 1768–71." *Notes and Records of the Royal Society of London* 49, no. 2 (July 1995): 245–260.

Chico, Tita. "Gimcrack's Legacy: Sex, Wealth and the Theater of Experimental Philosophy." *Comparative Drama* 42, no. 1 (Spring 2008): 29–49.

Clark, Andrew, ed. *"Brief Lives," Chiefly of Contemporaries, Set Down by John Aubrey.* 2 vols. Oxford, 1898.

Copeman, W. S. C. "Dr Jonathan Goddard, F.R.S. (1617–1675)." *Notes and Records of the Royal Society of London* 15 (July 1960): 69–77.

The Country Spy, or a Ramble Through London. London, 1730.

Derham, W. *Philosophical Experiments and Observations of the Late Eminent Dr. Robert Hooke.* London, 1706.

'Espinasse, Margaret. *Robert Hooke.* London, 1956.

Evelyn, John. *A Panegyric to Charles the Second.* London, 1661.

———. *Silva, Or a Discourse of Forest-Trees*, 4th ed. London, 1706.

———. *The Diary of John Evelyn.* Ed. William Bray. 2 vols. London, 1950.

Fogg, G. E. "The Royal Society and the Antarctic." *Notes and Records of the Royal Society of London* 54, no. 1 (January 2000): 85–98.

Foster, M. "A Note on the History of the Statutes of the Society." *Proceedings of the Royal Society of London* 50 (1891–1892): 501–515.

Glanvill, Joseph. *Plus Ultra: Or, the Progress and Advancement of Knowledge Since the Days of Aristotle.* London, 1668.

Granville, Augustus Bozzi ["One of the 687 F.R.S.---sss"]. *Science Without a Head; or, The Royal Society Dissected.* London, 1830.

Grew, Nehemiah. *Musæum Regalis Societatis.* London, 1681.

Bibliography

Hall, Marie Boas. *Promoting Experimental Learning: Experiment and the Royal Society, 1660–1727*. Cambridge, 1991.

Hatton, Edward. *A New View of London*. 2 vols. London, 1708.

Hemmen, George E. "Royal Society Expeditions in the Second Half of the Twentieth Century." Supplement, *Notes and Records of the Royal Society of London* 54 (September 20, 2010): S89–S99.

Henry, John. "Wilkins, John (1614–1672)." In *Oxford Dictionary of National Biography*. Oxford, 2004; online ed., https://doi.org/10.1093/ref:odnb/29421.

Hill, Sir John. *A Dissertation on Royal Societies*. London, 1750.

———. *A Review of the Works of the Royal Society of London*, 2nd ed. London, 1780.

Hooke, Robert. *Micrographia: Or Some Physiological Descriptions of Minute Bodies*. London, 1665.

———. *The Diary of Robert Hooke*. Ed. Henry W. Robinson and Walter Adams. London, 1935.

Hornsby, Thomas. "On the Transit of Venus in 1769." *Philosophical Transactions* 55 (1765): 326–344.

Hubert, Robert. *A Catalogue of Many Natural Rarities*. London, 1665.

Hughes, J. "Divine Right or Democracy? The Royal Society 'Revolt' of 1935." Supplement, *Notes and Records of the Royal Society of London* 64 (September 20, 2010): S101–S117.

Hunter, Michael. "The Social Basis and Changing Fortunes of an Early Scientific Institution: An Analysis of the Membership of the Royal Society, 1660–1685." *Notes and Records of the Royal Society of London* 31, no. 1 (July 1976): 9–114.

———. *Establishing the New Science: The Experience of the Early Royal Society*. Woodbridge, UK, 1989.

Bibliography

Kronick, David A. "Notes on the Printing History of the Early 'Philosophical Transactions.'" *Libraries and Culture* 25, no. 2 (Spring 1990): 243–268.

Lloyd, Claude. "Shadwell and the Virtuosi." *PMLA* 44, no. 2 (June 1929): 472–494.

Lyons, Sir Henry. *The Royal Society, 1660–1940: A History of Its Administration Under Its Charters.* Cambridge, 1944.

Macky, John. *A Journey Through England, in Familiar Letters from a Gentleman Here, to His Friend Abroad.* London, 1714.

Martin, D. C. "Former Homes of the Royal Society." *Notes and Records of the Royal Society of London* 22, nos. 1–2 (September 1967): 12–19.

Mason, Joan. "Hertha Ayrton (1854–1923) and the Admission of Women to the Royal Society of London." *Notes and Records of the Royal Society of London* 45, no. 2 (July 1991): 201–220.

———. "The Women Fellows' Jubilee." *Notes and Records of the Royal Society of London* 49, no. 1 (January 1995): 125–140.

Miller, David Philip. "The Usefulness of Natural Philosophy: The Royal Society and the Culture of Practical Utility in the Later Eighteenth Century." *British Journal for the History of Science* 32, no. 2 (June 1999): 185–201.

Miller, Henry Knight. "Henry Fielding's Satire on the Royal Society." *Studies in Philology* 57, no. 1 (January 1960): 72–86.

Newes from the Dead, or a True and Exact Narration of the Miraculous Deliverance of Anne Greene. Oxford, 1651.

Oldenburg, Henry. *Correspondence.* Ed. and trans. A. Rupert Hall and Marie Boas Hall. 13 vols. London, 1965–1986.

Olson, R. C. "Swift's Use of the 'Philosophical Transactions' in Section V of 'A Tale of a Tub.'" *Studies in Philology* 49, no. 3 (July 1952): 459–467.

Bibliography

Pepys, Samuel. *The Diary of Samuel Pepys*. Ed. Robert Latham and William Matthews. 11 vols. London, 2000.

Philosophical Collections. 7 issues. London, 1679–1682.

Philosophical Transactions. Online at Royal Society Publishing, http://rstl.royalsocietypublishing.org.

Pope, Walter. *The Life of Seth, Lord Bishop of Salisbury*. Oxford, 1961.

Robinson, H. W. "An Unpublished Letter of Dr Seth Ward Relating to the Early Meetings of the Oxford Philosophical Society." *Notes and Records of the Royal Society of London* 7, no. 1 (December 1949): 68–70.

Selderhuis, Herman J., ed. *The Calvin Handbook*. Grand Rapids, MI, 2009.

Shadwell, Thomas. *The Virtuoso*. Ed. Marjorie Hope Nicolson and David Stuart Rodes. Lincoln, NE, 1966.

Sorbière, Samuel de. *A Voyage to England, Containing Many Things Relating to the State of Learning, Religion, and Other Curiosities of That Kingdom*. London, 1709.

South, Sir James. *Charges Against the President and Councils of the Royal Society*, 2nd ed. London, 1830.

South, Robert. *Discourses on Various Subjects and Occasions*. Boston, 1827.

———. *Sermons Preached upon Several Occasions*. 4 vols. Philadelphia, 1844.

Sprat, Thomas. *The History of the Royal-Society of London*. London, 1667.

Stewart, Larry. "Other Centres of Calculation, or, Where the Royal Society Didn't Count: Commerce, Coffee-Houses and Natural Philosophy in Early Modern London." *British Journal for the History of Science* 32, no. 2 (June 1999): 133–153.

Syfret, R. H. "Some Early Critics of the Royal Society." *Notes and Records of the Royal Society of London* 8, no. 1 (October 1950): 20–64.

Bibliography

Thomas, Jennifer M. "A 'Philosophical Storehouse': The Life and Afterlife of the Royal Society's Repository." PhD diss., Queen Mary University of London, 2009.

Tinniswood, Adrian. *His Invention So Fertile: A Life of Christopher Wren*. London, 2001.

Turnbull, G. H. "Samuel Hartlib's Influence on the Early History of the Royal Society." *Notes and Records of the Royal Society of London* 10, no. 2 (April 1953): 101–130.

University of Oxford. *Oxford University Statutes*. Trans. G. R. M. Ward. 2 vols. London, 1845–1851.

Wallis, John. *A Defence of the Royal Society*. London, 1678.

Ward, Edward. *The London Spy Compleat*. London, 1703.

Ward, John. *The Lives of the Professors of Gresham College*. London, 1740.

Waters, David W. *The Art of Navigation in England in Elizabethan and Early Stuart Times*. London, 1958.

Wotton, William. *Reflections upon Ancient and Modern Learning*, 3rd ed. London, 1705.

Wren, Christopher. *Parentalia: or, Memoirs of the Family of the Wrens*. London, 1750.

Youngson, A. J. "Alexander Bruce, F.R.S., Second Earl of Kincardine (1629–1681)." *Notes and Records of the Royal Society of London* 15, no. 1 (July 1960): 251–258.

INDEX

Index

Index

Index

Index

Index

Index

Index

Index

Index

Index

ADRIAN TINNISWOOD is senior research fellow in history at the University of Buckingham and the author of several books, including *Behind the Throne: A Domestic History of the British Royal Household* and the *New York Times* bestseller *The Long Weekend: Life in the English Country House, 1918–1939*. He was awarded an OBE for services to heritage by Her Majesty, the Queen, and lives in Bath, England.